Miracles, Mystics, Mathematicians

Miracles, Mystics, Mathematicians: Searching for Deep Reality focuses on the lives and writings of some of history's most influential mathematicians and the impact that their mystical beliefs had on their lives and on their mathematical work. Modern biographers often cleanse the lives of renowned scientists of any hint of mysticism or occultism. Such threads are sometimes regarded as relics of the superstitious past: flaws that need to be hushed up, marginalized, or reinterpreted. This book represents a minor attempt to push back against this tendency and to examine these aspects of the history of mathematics with seriousness and intellectual curiosity.

Features

- A breadth of scope covering many centuries
- Suitable for anyone interested in mathematics, history, philosophy, paranormal phenomena, psi-research, mysticism, or in any combination of the above.
- An almost unique account of known histories, examined from a new vantage point

Sasho Kalajdzievski is a Senior Scholar in the Department of Mathematics at the University of Manitoba.

T0273779

Miracles, Mystics, Mathematicians

Searching for Deep Reality

Sasho Kalajdzievski

CRC Press is an imprint of the
Taylor & Francis Group, an **informa** business

AN A K PETERS BOOK

First edition published 2024
by CRC Press
6000 Broken Sound Parkway NW, Suite 300, Boca Raton, FL 33487-2742

and by CRC Press
4 Park Square, Milton Park, Abingdon, Oxon, OX14 4RN

CRC Press is an imprint of Taylor & Francis Group, LLC
© 2024 Sasho Kalajdzievski

ISBN: 9781032252308 (hbk)
ISBN: 9781032251325 (pbk)
ISBN: 9781003282198 (ebk)

DOI: 10.1201/ 9781003282198

Typeset in Minion
by Deanta Global Publishing Services, Chennai, India

To Nina, Darja, Damjan, Timjan, Mia, Ada, and Leo; with love.
За Нина, Дарја, Дамјан, Тимјан, Миа, Ада и Лео; со љубов.

Contents

Prologue and Acknowledgements

I N 2009 I SUBMITTED an eight-page article titled *The Metaphysics of Some Mathematicians: Short Notes* to the Mathematical Intelligencer. It was rejected, and I have never submitted it elsewhere. The editor of the Intelligencer at the time was the prominent mathematician, writer, and political activist Chandler Davis. I knew him personally: he was the very first person who directed my wife and me academically upon our arrival in Canada.

The article turned out to be the beginning of a larger project. It also led to a meaningful exchange of lengthy emails between the two of us, regarding so-called paranormal phenomena. Professor Davis was both judiciously critical and open minded. The whole affair discretely steered my interests, and I started to pay more attention to mystic and paranormal events related to the lives of eminent mathematicians. It turned out to be the beginning of a larger project.

In 2015, six years later, I had a clear idea for the content of an expanded, book-sized project, and after I had completed the early versions of the first three chapters, I started looking for a publisher.

This proved to be a rather dreadful affair. The main problem was that the project was uncategorizable: on one hand, there was just too much (history of) mathematics in it to be acceptable to publishers specializing in paranormal phenomena, and, on the other hand, the paranormal phenomena that were included were mostly considered antithetical to the views of the scientific establishment. I reckoned that the only chance, however small it might have been, was to find a publisher among the scientifically oriented ones.

So, I started the search. Most of the editors unceremoniously rejected the project. In a few cases the project – an early version of the first two or three chapters – was given to reviewers for evaluation. Some of the reviewers seemed to have been rather irritated by the audacity of someone submitting such a proposal. One of them directed me to Shambhala, a publishing house specializing in 'Buddhism, Mindfulness, Wisdom Traditions, Health, Yoga, and more' and the legendary underground kingdom, assumed by many to have no 'real' existence. Since we do not cover the topics in which Shambhala publisher specializes, it seems that it was the second meaning – the supposedly nonexistent, underground kingdom Shambala – that applied, and thus the reviewer was in effect telling me to get lost.

It took eight years and many submissions to finally find the *right* editor: Callum Fraser from CRC Press/Taylor & Francis Group. He not only managed to drive the project past the editorial board, the barrier that the other editors were rather reluctant to challenge, but was also very supportive in many different ways. I am very thankful to him.

Also thanks to the copyeditor Louisa McDonnell for her excellent work.

Another person whose involvement was crucial is Snežana Lawrence. There would have been no chance at all for the project to be accepted if it weren't for her strongly positive review of an initial part of the project. Thank you!

I also thank Fereidoun Ghahramani and Ghollamhossein Iraghi Moghaddam for their (independent) translations from Farsi into English of a quatrain from Rubaiyat by Omar Khayyām.

Special thanks and my deep gratitude to Darja Barr, my 'in-home' editor, who found time and energy to read and scrutinize the entire manuscript. Her input was invaluable to me. Thank you, Dar!

Preface

THE MAIN THEME OF our story focuses on the lives and writings of some influential mathematicians, on the impact of their mystical beliefs on their lives, and on their mathematical work. Less prominent mathematicians will also be included, especially those who played major roles as mystic philosophers, mystic experiencers, or paranormal researchers. The choices we make are subjective; this is not a comprehensive anthology.

I will take the metaphysical ideas of our heroes, as well as their mystic experiences, at least at face value. At times, and with increasing frequency as we go, we will play the role of 'devil's advocate' and will provide scientific evidence supporting the genuineness of these experiences and the viability of the associated mystic/paranormal theories. We will mainly follow the dictum of experimental sciences and, somewhat reluctantly, avoid relying on anecdotal evidence.

We will not evaluate mystic beliefs and experiences from the materialistic viewpoint, sometimes misleadingly called the 'Western rationalistic' viewpoint, for that would be tantamount to declaring most of the mathematicians we will encounter as being at least proto-schizophrenics. The ideas of these ingenious people are typically far off of the orthodoxy grid, and they require a special perspective in order to assess them properly. It would be fair to take these individuals seriously, if nothing else but because of their intellectual eminence. Ingenuity is incompatible with conventionality.

No belief or mystical event will be dismissed out of hand, even in cases where it is viewed as being too outrageous or even ridiculous by the scientific establishment. Everything experienced or believed by our mathematicians will be considered seriously. 'Everything' means everything: from telepathy and clairvoyance, through spirits and post-mortem existence, to alchemy and astrology. The issues we will encounter may not be resolvable through compromises and the search for a middle ground.

Posturing objectiveness and conforming to theoretical dogmas certified by the Ministry of Scientific Truth do not attract me. We will defy the Ministry and go where the roads take us.

According to many studies almost all of us have had or will have some mystic-like experience. Interpreting these events is another matter. We have been conditioned, even brainwashed, to accept as reasonable only the explanations that conform to the materialistic paradigm, starting with the thesis that everything is sourced in the *objective* material world. If you see an apparition of uncle Joe then, we are told, the only sane explanation is that it was an emanation of your brain. If it happened that uncle Joe was killed in a traffic accident far away from you and at exactly the same time, then we are told it was a coincidence. Since such cases were published hundreds of times in books and articles, we also have to swallow the claim that all of these cases are just a bunch of coincidences. The most obvious explanation, that uncle Joe had something to do with the production of his own apparition, is not to be considered seriously for, as we are told, that would make us gullible believers, followers of pseudo-science, naïve, uneducated, and a ton of other unflattering epithets.

We will make a clear distinction between the mystical and religious components of their lives, and we will rarely deal with the latter. For example, the fact that Ramanujan was a devout Hindu plays a role in our narrative insofar as his Hinduism is reflected in his mystic experiences, including his perceived direct communion with certain deities.

Sometimes modern biographers tend to cleanse the lives of renowned scientists of mysticism and of anything that has traces of occultism. Such threads are considered to be relics of the superstitious past, flaws that need to be hushed up, marginalized, or reinterpreted. This project is a minor attempt to correct this distortion.

It is important to realize that so-called paranormal or mystic-like phenomena are not incompatible with the great achievements of our sciences. It is the associated basic postulates and the theories generated by them that cannot coexist. The distinction is often overlooked by skeptics who either ignore or a priori dismiss the large and expanding body of scientific experiments producing results that contradict the basic axioms of the materialistic paradigm.

The narrative will go anti-chronologically most of the time. This is forced. The alternative would be to start with the fantastic Pythagorean reality and end with the secretive and subdued personal world of Gödel.

That would be anticlimactic. The exception to the anti-chronological order will primarily be due to the complexity of the phenomena. Thus, for example, Poincaré will precede Brouwer, though the latter was born some two decades before the former.

In light of the broadness of our topics, we are forced to substantially limit the depth of our narrative. This concerns the mystic ('miraculous'), mathematical and biographical parts of the story. Regarding the mystic or paranormal aspects, we will support them with a sample of scientific, laboratory results; we mention again that we will avoid anecdotal evidence despite the extraordinary profusion of such cases.

It was especially disagreeable to me that I had to reduce the biographies of the people we will consider to bare outlines: after reading extensively about their lives I had a strong impression, especially in some cases, that I knew them personally. I felt empathy for their successes and failures, happiness and sadness, joys and tragedies. So I feel that I am committing a kind of sacrilege by reducing their whole lives to résumés that fit a page or two. I apologize for that, but I had no choice.

The mathematics that we will cover will be watered down, and it will mostly be included to illustrate the significance of the research of the respective mathematician. What we choose to present will also depend on how interesting and how accessible the associated selection is, and this is always a rather subjective call. Consequently, the amount of mathematics that will be associated with a particular mathematician in our narrative will not necessarily be proportional to his/her relative mathematical eminence.

I am not a philosopher. I may or may not like the philosophical ideas and theories; I certainly dislike the prevalent philosophical verbosity that cloaks some of those theories.

I do not have aspirations to present cold historical accounts. The reasons are simple:

(a) I am not a historian; in fact, I couldn't be one. History is based on the assumption that time is linear, which I don't accept.

and

(b) I cannot distance myself from the lives of people and view them from afar. I cannot forfeit my own views or interpretations, and I cannot vanquish my own emotions. In that respect, this is a book about its author too. It is always so; I merely admit it.

However, I will not willfully pollute facts with fiction. The sources will be given and referenced, except in a few cases, which will be clearly indicated. It is then up to the reader to decide if the anecdotes from the lives of mathematicians that we convey are authentic. For me, the credibility of specific sources, or the believability of specific events, is not of major significance. One can always dismiss anything as fiction, myth, delusion, hallucination, or deception. By 'anything', I do mean 'anything', and that includes the material world in particular. The overall message conveyed by our heroes and by their lives is much more important than the particularities of some cases. This message may be very roughly approximated by the following paraphrase of a lament by Pascal:

> The reality beyond our physical perceptions is a thing so important that only those who have lost all feeling can rest indifferent to it.[1]

Here we go! Along the way, we will burn the bridges behind us. With some pleasure.

---

[1] Pascal's original quote is: *The immortality of the soul is a thing so important that only those who have lost all feeling can rest indifferent to it.* (Pensées (Thoughts))

CHAPTER 1

On Defining Mystics, Miracles, and Mathematicians

What if you slept, and what if in your sleep you dreamed,

And what if in your dream you went to heaven,

And then plucked a strange and beautiful flower,

And what if when you awoke

You had the flower in your hand.

Ah!

What then!

SAMUEL TAYLOR COLERIDGE

1.1 A MYSTIC EVENT OR A HALLUCINATION?

You are falling asleep. You casually open your eyes one more time and notice a silhouette of a man. You are now fully awake, and your sight has cleared. You plainly see a man staring at you. Then it strikes you that something very strange is going on. As your heart starts pumping faster, the apparition dissolves. You stand up, put the light on, and check the room. Nothing! 'What on earth has just happened?' you wonder.

DOI: 10.1201/9781003282198-1

We just described a generic case of a hypnagogic hallucination. It happens in the state on the boundary between sleeping and waking, just before falling asleep. At the other end of the night, just before one wakes up, lies the domain of hypnopompic hallucinations. These two, especially the former, are common types of hallucinations, and many studies show that the majority of people experience them at least once in a lifetime. There is no controversy here: it is widely accepted that the phenomenon is ubiquitous and genuine.

The problem starts at the level of interpreting such an event. There are clearly two possibilities: it is either a deception played upon you by your subconscious that has no base in objective reality, or it is something that emanates from sources external to your psyche. In the latter case 'a hallucination' is usually called 'a vision' or 'an apparition'. In both cases, the terms 'objective reality' and 'external' have no more than intuitive meanings.

Before we review the arguments supporting each of the two interpretations, we consider the following genuine anecdote. Ivan[1] was a mathematician who lived in a so-called socialist country. At the time of the event that he recounted to me he was in his 20s, and he did not know, nor cared about, virtually anything that was spiritually oriented. In fact, the word 'spiritual' was meaningless to him, or, at best, it was associated with young ladies who read and wrote poetry. The only religion he had some knowledge of was dialectic materialism – and even that did not interest him; he accepted it tacitly only in order to be aligned with the local social flow. Ivan lived alone – the times were not so bad – in a one-half-bedroom apartment in a typical working-class high-rise neighborhood. One ordinary night, after an ordinary day, he went to bed, relaxed for a while, and fell asleep. He does not know when exactly, but it must have been shortly after he fell asleep, something made him open his eyes. To his utter astonishment, he discovered he was not alone: there was a figure of a man by his bed. After a few seconds of dumbfounded staring, his bewilderment escalated, for the man he saw was crucified on a large cross apparently erected at the foot of his bed. That much he knew: Jesus Christ! The scene was illuminated, as one would expect it to be. Ivan stared unbelievingly for – what he thought afterwards – a long time. He could not make any sense of what he was seeing. He closed his eyes, and opened them again: the apparition was still there. He moved his head and the scene changed

---

[1] Not his real name.

into a three-dimensional perspective: it was not a projection on the opposite wall. Eventually, Ivan gathered his courage, stood up, and took a step in the direction of the vision, at which point the cross, Jesus and all, dissipated into thin air.

Visions of Jesus Christ have been reported thousands of times by various mystics, as well as by uninitiated ordinary people. After a similar experience, Pascal, as we will see, cried tears of joy. Ivan did not cry. He deliberated over his episode the next day, called it a hallucination that had no basis in objective reality, and went on with his regular life. According to modern orthodox psychology, he proved his mental sanity, for even considering another explanation is deemed symptomatic of some mental deficiency. So Ivan's choice was a pragmatic step without any doubt, for the alternative interpretation, combined with frankness, may have led him straight to the lunatic asylum.

From a psychological point of view, there is no difference between visions, apparitions, and hallucinations: they are all hallucinations. Modern psychology generously refrains from diagnosing you as a hysteric if you have experienced one since that would be tantamount to declaring most of humanity as consisting of nutcases. The normalcy test is pushed one step further, at the stage of interpreting events. If a person experiences a 'hallucination', then he is pronounced psychologically normal if he checks it against 'objective reality' and then concludes that it is an emanation of his subconscious. If this algorithm is not followed and one considers the *wrong* interpretations seriously, psychology tells us, then such a person is to a degree a delusional hysteric and may be in the state of proto-schizophrenia.

The code word here is 'reality'. Psychology, of course, never bothers with defining or explaining this momentous notion. It is merely tacitly assumed that the objective world, as perceived through our senses, is external to the perceivers and thereby gives a good reference point with respect to which we can judge what is normal and what is not. As a consequence of this casualness, their definition of a normal attitude with regard to hallucinations is rather tentative. Mathematicians are instinctively aware of the weakness of such arguments. One does not have to know quantum mechanics in order to doubt the absoluteness of matter.

Interpreting a mental experience is the first basic act that distinguishes a mystic from a 'normal' person. A mystic accepts the thesis that, through visions, one accesses a kind of objective reality and that one's mind is not a creator of these experiences but rather acts as a portal through which one can arrive at deeper, transcendental states. As a corollary, mystics accept the claim that there is an objective reality beyond the material world and, moreover, that this reality is accessible. It is precisely this thesis of accessibility of the deeper or divine reality beyond bodily perceptions that distinguishes a mystic from a typical religious follower. However, our definition of a mystic – to be discussed in the following sections – will be broader.

It follows that, from the point of view of psychopathology, mystics are at least delusional. However, this view is very simplistic, and we are not going to follow it.

The ultimate source of the differences in interpreting hallucinations lies in the choice of the axioms that give rise to various theories or paradigms. Whether one would call an experience a hallucination or a mystic experience depends not on the experience per se but rather on interpreting the origin of the corresponding event, which is in turn sourced in the choice of the acceptable axioms. The psychological categorical choice of the brain as the only source of hallucination is no more than an axiom. It is a statement taken to be true without any proof whatsoever; it is merely compatible with other entrenched materialistic axioms. If you happen to be a mathematician who is among the majority who has experienced some kind of hallucination, then you would be inclined to consider other logical explanations on equal footing, irrespective of the prevailing dogmas. This opens doors to the no-go land of taboo theories.

1.2 WHAT IS MYSTICISM?[2]

On May 6, 1793, by a decree of the Russian Empress Catherine the Great, Stepan Lubovedsky, a Polish general and a brave warrior, was awarded the Order of St. Alexander Nevsky. Thus started the Lubovedsky nobility. Some 35 years later, Prince Yuri Lubovedsky was born. The decades after the Napoleonic wars were a time of relative peace within the vast Russian Empire, and the young prince had a happy and comfortable childhood. About a quarter of a century later, his good life climaxed with his marriage to a beautiful young woman whom he deeply loved. Soon they were looking forward to the crown of their love, a new baby. And here the story[3] takes an abrupt turn.

The princess and the baby died during childbirth. Prince Lubovedsky's life crumbled into grief and despair. He was unable to accept the loss and start anew. For the first few following years, he indulged in reading

---

[2] Since we are using 'mysticism' as a criterion in this book, defining it seems important at this stage.
[3] Warning: longish story ahead! It is an exception; I prefer short stories.

occult books and then tried desperately to contact the spirit of his wife. That failed. Then he crisscrossed the world for decades searching for clues and for wisdom that would help him understand his personal tragedy and the meaning of life in general. He checked European libraries, traveled to Egypt, India, Tibet, and many other places in between, and met many a wise man. He was already in his 60s when he found himself in Kabul, Afghanistan, worn out and disillusioned.

While in Kabul he visited his good friend Aga Khan, who introduced Prince Lubovedsky to another visitor: a very old, scrawny Tamil man. The prince felt attracted to this old man, who, as it turned out, had visited Russia and spoke Russian. Eventually, they went together to a café and chatted for a long time. One moment during the conversation, the old Tamil man declared that what they had been speaking about so far were trifles that missed the point of their meeting, and he simply stopped talking. A long silence ensued. We quote the prince:

> I said to myself, 'Poor fellow, doubtless his thinking faculty has already began to weaken with age and his mind has began to wander'. I felt painfully sorry for this dear old man. [] I reflected that my mind also would soon begin to wander and that the day was not far off when I too would not be able to direct my thoughts, and so on. I was so lost in these heavy but fleeting thoughts that I even forgot the old man. Suddenly I again heard his voice. The words he spoke instantly dispelled my gloomy thoughts, and shook me out of my state. My pity changed to such an astonishment as I think I had never experienced in my life before.'

> 'Eh, Gogo, Gogo! Forty-five years you have worked, suffered and labored incessantly, and not once did you decide for yourself or know how to work so that, if only for a few months, the desire of your mind should become the desire of your heart. If you had been able to attain this, you would not now in your old age be in such solitude as you are!'

> The name 'Gogo' which he had used made me stare with amazement. How could this Hindu, who saw me for the first time somewhere in Central Asia, know the nickname by which I had been called in my childhood sixty years before, and then only by my mother and nurse, and which no one had ever repeated since then? Can you imagine my astonishment?

The old man, who, according to this account, was somehow able to access the life story of the prince, then offered to help him, under the condition that the prince agreed to 'consciously die to the lie [that the prince] had led until [then]'.

Prince Lubovedsky ended up in Sufi monasteries (tekkes) somewhere close to the border between Tajikistan and Tibet, where he devoted the last five years of his life to contemplation and meditation. This story comes from the intriguing late 19th-century travelogue *Meetings with Remarkable Men* written by George Gurdjieff.[4] Gurdjieff himself was an influential mystic; however, he was not interested in mathematics and so we will leave his life story aside.

There are a few aspects of this story that we will use as a pretext to introduce some of the main features of our narrative. First, there is the 'supernatural' power of the old sage to access long-forgotten memories of someone quietly drinking coffee with him. This can be categorized either as a mini miracle or as fiction. Followers of scientism and other dogmatic materialists scoff at anything that is incompatible with their beliefs; 'miracles', mini or maxi, are most certainly included in the 'to-scoff' list. I do not buy the materialistic paradigm anymore, and, consequently, we will keep the doors wide open for other interpretations and for other theories in our narrative.

There is one more feature of the above anecdote that we will consider more thoroughly: it is the old man's statement about synchronizing the desires of mind and heart. How does one achieve harmony between the desires of mind and of heart? And what does that mean?

In the story about the prince and the sage, the latter implied that the paths leading to knowledge and to the attainment of harmony of 'mind and heart' are inward bound. *Mystics*, by definition, are those who have traveled that road, and, moreover, they regard their experiences along the mystic paths as emanating from a source not solely confined to their own consciousness. *True mystics* have gone all the way. Some of them have done it in one instant, others took a long time. The instantaneous mystics get there involuntarily and seemingly during unpredictable spontaneous flashes of illumination. The adept travel inwards patiently and deliberately.

This inward travel, as the mystics tell us, is usually achieved by suppressing the rational mind. For example, we are told that Padmasambhava, the 8th-century Buddhist mystic, went two years without thinking as a part

---

[4] [Gur].

of his initiation. Not thinking is one mode of depriving the senses of perceptual inputs and, at the same time, denying one's selfhood. The latter is also a prerequisite for the deliberate attainment of deep mystical states. The famed Sufi poet Rumi expressed it succinctly in the following stanza:

> I have found true being in non-being
> So I wove my soulhood into nothingness.

The acknowledgment of the empirical existence of a way inwards is one of the main characteristics of mysticism. At the end of this journey, true mystics tell us, there is direct, experiential awareness of the presence of the Ultimate Consciousness, Divine Reality, or God. This is, then, the true understanding of reality, 'which transcends the temporal categories of the understanding, relying on speculative reason.'[5]

True mystics reach ecstatic states of consciousness, achieving a sense of unity with (what they perceive as) the deeper, timeless reality. For them, mysticism is not a philosophical scheme but rather a way to experimental spiritual wisdom.

Most of the mystic travelers only reach partway. They will constitute our first extension of the set of people described as *mystics*: we will now include those who have mystical-like experiences and who do not treat these experiences as being products only of their own subconscious. What a 'mystical-like experience' means is open to interpretations. Consider again hallucinations. They are automatically mystical-like experiences. However, if the experiencer considers their source to be external to his or her psyche, then they become full-blown mystical experiences. The whole mystic realm is in the eye of the beholder.

Hallucinations are ubiquitous phenomena: as we have mentioned, according to many studies, the majority of people have at least one hallucination during their lives. Among hallucinations, there are various types, and of various degrees of realism – elaborating this further is out of the scope of this book. What is important to us is the attitude of societies towards people who report hallucinations, specifically towards the subset of experiencers who believe in the objective reality of what they have seemingly perceived. Today they would possibly be diagnosed as pre-schizophrenics, or even as psychotic (insane). The same people in other times might have been called anything between heretical and saintly.

---

[5] [Hap], p. 37.

Finally, we mention what we will call *rational mystics*. These are people who have not had or do not admit to having had a personal mystical experience, yet accept the authenticity of mystic phenomena. Their acceptance may be based on what they consider to be the available anecdotal or experimental evidence. For example, this class includes those who accept the scientific evidence of the genuineness of clairvoyance. Clairvoyance presumes the existence and accessibility of transcendental realms out of the reach of common perception. From accepting the clairvoyance phenomena to accepting the realness of the mystic travels there is only one simple step.

With this last definition, we further extend the class of mystics, which will now include rationalists sympathetic to the idea of mysticism. We are forced to take this step: as indicated above, during the last two centuries or more, it was risky to one's well-being to declare openly mystic-like experiences and, moreover, to accept them as emanating from the objective reality. It was considerably less dangerous to consider these cases as conditionally or philosophically acceptable. As a consequence of our inclusive definition, we will see a few relatively modern mathematicians who presented themselves as rational mystics and who, as it seems, may have had mystical experiences that set in motion changes in their life, or at least in their philosophy, but were reluctant to make them public.

In order to clarify any misunderstanding, we note again that being religious is not necessarily related to being mystic, though these two classes are far from being disjoint. Mystic orders have at times been regarded as 'heretic' threats to the established religions, even though some religions, including early Christianity, were based on mystical doctrines, such as the possibility of comprehending a deeper reality, or God, through devotion, self-denial, and various mental exercises. In this book, we will strictly separate mysticism from religiousness. The latter will be considered insofar as it is related to the lives of our mystics-mathematicians. The former will be our main focus of interest. In particular, mystic-leaning schools, such as Sufism and Rosicrucianism, will be mentioned as being linked to the lives of some of the mathematicians we will meet.

Since this is not a treatise in mysticism, what we define as mysticism will be primarily used as a provisory criterion to select our mathematicians, not a thesis that needs to be defended.

1.3 WHAT IS A MIRACLE?

Now that our introduction of mysticism is over with, defining the other two notions appearing in the title of this book is a much smoother ride.

In societies with a dominating paradigm of reality, a *miracle* is an event that is too far out of the frames of that paradigm and hence contradicts it. From the point of view of a strongly paradigmatic society, miracles do not happen at all, precisely because their possibility is not sanctioned by the dominating paradigm. The more established the paradigm, the less acceptable are any deviations from its doctrines, and the less such a society is tolerant of dissent. Centuries ago the *heretics* were burnt at the stake. In modern times, those who question the norms and boundaries of the sanctioned reality are often labeled as superstitious, delusional, or gullible.

There did exist paradigm-freer societies in human history. The concept of a miracle in societies with weak paradigms is almost meaningless: almost everything is a miracle and not a miracle at the same time. The depth of a societal paradigm is not necessarily related to the total knowledge of a society; it is a function of a type of recognized knowledge.

Miracles are always defined in reference to the state of cumulative knowledge of a society. This knowledge changes; hence the definition of miracle pertaining to a specific society changes too. Not always can this change be described as progress. For example, the modern scientific materialistic paradigm, as successful as it appears to be, rolled over and squashed many alternative reasonable models along its way. It was not always a quest to understand nature, but rather to dominate it.

When we deal with events designated as miracles, our point of reference will be the modern materialistic paradigm.

We are, of course, not claiming that everything unknown or not understood is a miracle. Miracles are not events that are at the moment out of reach of the acceptable models of reality but may be within their frames after some time of linear expansion of these models and the underlying theory. We define them to be events, or alleged events, that directly contradict the theoretical (implicit or explicit) axioms of the dominating paradigm. The boundary between the phenomena that outright contradict the dominating scientific paradigm, and those that are not explainable now but may be within a future scientific theory that will keep the basic dictums of the current one, is often very fuzzy.

Consider, for example, the case of telepathy encountered in the story of the prince and the sage. Telepathy by itself, to the extent that its genuineness is acceptable, is a weak miracle: it is out of the scope of our current science, but it can be viewed as something straight ahead on the road of its linear progress. The thesis that the sage accessed all of the memories of the prince's life is much more miraculous. How did the sage decode the life of the prince while the latter's brain waves were generated by neurons fired by thoughts that had nothing to do with his life story? From the point of view of materialism, such an event is not possible. No amount of anecdotes or experimental data could alter that conclusion, since paradigms are intrinsically immune to self-destruction.

Since our modern dominating paradigm of the global society of today will be our reference point when talking about miracles, we will stay with it for a moment. What is it anyway? Scientific materialistic rationalism may be one acceptable answer. What is its basis? The theoretical basis of the materialistic paradigm is determined by a few axioms that are seldom explicitly stated. One such axiom postulates the objectiveness of material reality.

It is not such a firm postulate as it used to be: micro-reality has been undermined by experimental quantum physics. But the ontological axioms in the basis of modern science have been firmly entrenched for centuries. Some are, for example, that matter precedes and generates consciousness and, related to it, that an individual human consciousness perishes with the demise of the body. There are many corollaries of these statements. One is the following: 'There are NO spirits which have no body but can communicate and influence the world.' Of course, this last quote is chosen on purpose. The point is that if we delete there the capitalized NO, we arrive at a verbatim statement made by Kurt Gödel, who, as we see and will see, vehemently disagreed with the materialistic axioms.

For the sake of Gödel, for the sake of the other mystic mathematicians we will encounter, and for our own sake, we will allow the possibility of alternative reasonable paradigms.

1.4 WHAT IS A MATHEMATICIAN?

A *mathematician* is someone who understands and appreciates the art of mathematics.

Understanding mathematics is not the same as knowing mathematics. Indeed, the field of mathematics expanded in so many directions that today it is humanly impossible to be familiar with all of it. By understanding mathematics, we mean having personal experience in doing some kind of research in mathematics, where, in this context, we use the term mathematics to stand for what is sometimes called pure mathematics.

As is the case with the notions of mystic and miracle, the meaning of the word 'mathematician' changed over time and over different societies. Up to the times of Leibniz, mathematicians were people who knew just about all of the mathematics that was known to historical humanity by their time.

Since the totality of our historical mathematical knowledge increased exponentially over time, it was more and more knowable as we go backwards in time. Hence, in our anti-chronological narrative, we will be encountering mathematicians who were all-around polymaths. Leibniz himself, besides conferring on us his mathematical legacy, covered a myriad of other endeavors; he was a diplomat, a philosopher, a chemist, a physicist, an alchemist… Johannes Kepler was not only a mathematician but also a famed astronomer. Isaac Newton, who was a professor of mathematics at Cambridge and who, together with Leibniz, was the founder of the notions of limit and derivative, is now remembered primarily as a physicist.

After Leibniz, mathematics developed in so many areas that it started to become inaccessible in its entirety to any individual, ingenious or not. Despite this trend for specialization, we will still utilize our initial definition of a mathematician. One does not have to have even a formal degree in order to fit in. Trying earnestly to discover mathematical puzzles, theorems, or theories makes one a mathematician according to our definition. We will stick with it almost all of the time; most of the exceptions will be a few minor mathematicians, but major mystics, such as, for example, Emanuel Swedenborg and John Dee.

CHAPTER 2

From Mathematics to Mysticism and Back

The true spirit of delight, the exaltation, the sense of being more than Man, which is the touchstone of the highest excellence, is to be found in mathematics as surely as in poetry.

BERTRAND RUSSELL, *MYSTICISM AND LOGIC*

2.1 FROM MATHEMATICS TO MYSTICISM

I was not looking for mathematicians in my survey of arcana of various sorts. They simply started coming my way. A mini-avalanche of them poured through the pages. They came as hermits and as philosophers; mystics and sorcerers; experimenters and experimental guinea pigs. They were experiencers of paranormal phenomena, and, in some extreme cases, they were themselves paranormal phenomena. I could not help but notice their profusion. Perhaps my own professional life predisposed me to pay attention to them.

Why are mathematicians more inclined to go the mystic way compared to other scientists? Why are they so disproportionately represented in the exploration of the world beyond our basic perceptions, and why are they so frequently protagonists in the field of (so-called) parapsychological research? This puzzled me for a while, but now I think I know the reasons. Counting myself as a mathematician, I believe I understand their peculiar way of thinking.

 DOI: 10.1201/9781003282198-2

Consider, for example, the following bizarre episode in the life of the American physician Cyrus Reed Teed. In 1869, Teed had a mystical experience during which a light-form slowly coagulated into the shape of a beautiful woman. The ensuing conversation between the two changed the flow of Teed's life: he switched his main vocation from practicing medicine to being a messianic cult leader and, along the way, changed his name from Cyrus to Koresh. Among the ideas he received from the beautiful apparition, the following left a profound impression on him: 'The universe', said the beauty, 'is a [] hollow globe [], and all life exists in its inner concave surface.'[1] Teed expanded on this theory in a book that he published soon after. The predictable reaction of the readership consisted mostly of pitying chuckles and ridicule.

The miraculous – or hallucinatory – connotations of the above story set aside, we take a look at the inverted-earth model of the universe. A genuine mathematician would never ridicule Teed's model to the advantage of the 'normal', convex-earth theory, for there is no mathematical difference between the two: just apply a simple inversion[2] with respect to a sphere of the usual model of three-dimensional space together with everything in it. Everything means everything, and that includes, for example, all of the canons that define reality according to standard science.

However, I am neither casting my vote for Teed's model nor am I endorsing the convex model of the 'universe'. That is beside the point. The argument I am making is altogether in a different direction. I am using this example to illustrate that typical genuine mathematicians are by trade, training, or perhaps by predisposition, very open-minded with regard to interpreting events and with regard to the world of ideas in general. If a model fits the data, then it is as good as any other, especially if it adheres through its simplicity to Occam's razor principle (stating that the best model is the simplest one – or the dumbest one).

Mathematicians are not guided by things that meet the eye but rather by things that meet the mind. They know very well that claiming that a statement is obvious, even postulating the convexity of the Earth, is potentially an ominous proposition and that it may either hide a fallacy or merely be an axiom leading to a theory not more valid than the one generated by

---

[1] [Stan], p. 145.

[2] We will not define it here. We will only mention that inversions send points outside the sphere to points inside the sphere, and vice versa; the center of the sphere is moved to 'infinity'.

the negation of the 'obvious' statement. The world of ideas is configured entirely differently from the world of basic perceptions.

Another important aspect unique to mathematicians, predisposing them to seriously consider the possibility of mystic realities, is that they do not have an axe to grind. Their professional work is completely immune from theories of the so-called material reality. Whether pigs can fly or not does not affect an iota of the veracity and meaning of the existing pure mathematical theories. This is not true for other natural sciences.

Mathematics immunizes its practitioners to any authoritarian proclamations. Prominent mathematicians may initiate theories and usher in modern streams of mathematical research. But they cannot deny a theorem even if a schoolboy discovers it. The only principle that guides genuine mathematicians is truth.

Further and notably, mathematicians are just about the only scientists who know what 'proof' means precisely. This matters in relation to our subjects of discussion. Every theory, including the one supporting the current materialistic paradigm, rests upon basic statements that are taken as being evident. These are axioms of a theory. We mention the axiom according to which matter emanates and supports consciousness; this is taken as obvious by the followers of the prevailing scientific paradigm. The alternative axioms, including the converse implication (consciousness generates matter), are fair game to any decent mathematicians, even at the price of being labeled superstitious and outdated. Mysticism is awash with alternative axioms.

We expand on the link between mysticism and the knowledge of formal theories further. There are many events that have the potential to be interpreted as mystic experiences. We mentioned already that the majority of people have experienced hallucinations at some points in their lives. Moreover, about 10% of humanity[3] has had at least one out-of-body experience (OBE). Further, about 4% have had a near-death experience following cardiac arrest.[4] And then, there are thousands upon thousands (millions upon millions?) of cases of premonition, precognition, telepathy, extraordinary coincidences, extraordinary experiences, and much more. Let us call these proto-mystic cases. Almost every person who has experienced a proto-mystic case would want to interpret that experience. If one is a mathematician, then the reaction is, in general, slightly different, compared to

---

[3] [Cha], p. 38.
[4] [Cha], p. 3.

the reaction of an average non-mathematician. Mathematicians know what models are: they know that each model has an underlying theory and that each theory is based on axioms. Following an ostensibly paranormal experience, the question of interpretation becomes a query on the axioms of so-called objective reality. It seems from the small sample that there are very few mathematicians who, after experiencing a proto-mystic event, automatically accept the orthodox interpretation; most of them would weigh the various theories and then decide. It is precisely this step that opens Pandora's box.

One by-product of the link between knowing what a theory means exactly, and keeping an open mind for unorthodox theories, including theories that encompass mysticism, is the rather conspicuous presence of some preeminent logicians in the ranks of mystics of various sorts. Among these, we will meet such highly distinguished logicians as Gödel, Turing, Brouwer, and De Morgan. These are not merely renowned mathematicians: they are some of the most influential scientists ever.

There is also another subtle lead from mathematics to mysticism. Mathematical theories in general establish links between seemingly unrelated ideas. One such good classical example is Galois theory, where we see a direct connection between ostensibly completely unrelated concepts: the ancient problem of doubling the volume of a cube and the theory of abstract algebra (field theory and group theory). Unusual phenomena that are often ignored by zealous followers of specific paradigms or teachings are precisely the incongruent singularities that mathematicians are trained to treat with the utmost seriousness. Hence, mathematicians with philosophical inclinations – just about everyone we will encounter – are often inclined to incorporate these far-off phenomena into their philosophical schemes, which immediately brings about mystic flavor in their theories. Leibniz's monadology is one such example, as we will see.

Finally, consider what it takes to set oneself along mystic ways. A deliberate mystic initiation requires achieving, what psychologists call, an altered state of consciousness (renamed by psychologist Stanislav Grof as a holistic state of consciousness). In a nutshell, one needs to put some distance between one's consciousness and the influx of percepts coming from the mundane world. This is precisely what it takes to do mathematical research. One is out of this world when *mathwalking*, thus becoming the proverbial absent-minded math professor. A good friend of mine slammed his bike into the rear window of a parked car, hurting himself and changing the shape of the front wheel into an 8, precisely because he was violating the basic rule of safe biking or driving, by thinking math at the same time. The flabbergasted owner of the car asked if the unusual biker was on drugs – which he certainly was not. This is not a singular case. After all, Brouwer was killed when he absentmindedly crossed a street. Clearly, the mystic states of consciousness are at levels different from the math states of consciousness. However, in the latter case, the seed is there.

The Platonic realm of ideas is intrinsically non-materialistic and esoteric. One may go as far as to postulate that ideas per se are spiritual manifestations. Pure mathematics – which I subjectively consider intellectually superior to any natural or applied science, including applied mathematics – lies squarely within the universe of pure ideas. Consequently, as we mentioned above, (pure) mathematicians tend to perceive deeper reality through their minds, not through their eyes. This only defines them as being in a permanent state of proto-mysticism. Going one step further is then just a matter of letting this proto-mysticism unravel.

Going mystic is the ultimate denial of the established paradigm. It takes an absolutely unpolluted and open mind to go headfirst against paradigms entrenched for hundreds of years. It takes mathematical knowledge to clearly discern the axioms of a belief, scientific or not, and to consider the possibility of a different reality. Mystics go in the same direction, except that their method is experimental and experiential, with a view to acquiring spiritual knowledge. Ultimately, the open-minded theories based on new axioms and mystical knowledge of the same axioms converge in the philosophies of the mathematicians we will encounter.

I am, of course, not suggesting that all mathematicians fit into one character mold. I am only saying that they are predisposed to view the world in its naked state, composed of transcendental ideas, independent of the prevalent paradigms of the day. The specific life stories, mystic beliefs, and the interconnection between these and mathematics will be discussed case-by-case for every individual mystic-mathematician that we will encounter.

2.2 FROM MYSTICISM TO MATHEMATICS

Spontaneous mystic episodes, or transcendental experiences, invariably affect one's life philosophy. In the case of true mystic experiences – when the mystic journey takes the traveler far within – the impact is almost always momentous. Weaker episodes are sometimes outwardly dismissed or marginalized. However, even in such cases, the seed of doubt is often deeply planted. Although the intensity of the subjective experience may be the most important factor, the final outcome is also influenced by other factors, such as, for example, the personality and the life of the experiencer, the ambient society, and his or her education. The causal pattern is rather complicated in the most general case.

However, the effect of mysticism or mystic experiences on mathematicians and their mathematics seems to be more tractable. Our sample of mathematicians is not very large, but, nevertheless, some order is visible. We will try to briefly categorize them in this section, and we will come back to this topic as we discuss the lives of individual mathematicians.

The most apparent general pattern seems to be the following: the later in someone's mathematical career a mystic event happened, the more radical the change of his[5] lifestyle and the greater the chances of complete abandonment of research mathematics. Late mystic bloomers will include Grothendieck, Gödel, Pascal (in a way), and perhaps De Morgan. Ramanujan is a prime example of a mathematician who started journeying along mystic roads relatively young; Brouwer and Cardano belong to the same class. In such cases, the mystically inclined mathematicians stay on course and continue their mathematical work without fundamental changes.

As we go backwards in time, this classification becomes less sharp. There are two main reasons: first, we know less about the lives of people who lived long ago, and second, popular acceptance of the reality of mystic experiences, and their ubiquity, was much more widespread in the past, especially in the Western world.

The congruity between mystical experiences and mathematical work seems to have been more pronounced in religious societies that tolerated discreet dissent, or in societies dominated by religions that shared the thesis with mysticism that the Divine Reality, or God, is apprehensible. In such circumstances, a mystic experience was acceptable; it might have been regarded as exceptional, but not as impossible, and most certainly not as a manifestation of some mental disease. In this context, mysticism and mathematics were often seen as converging or overlapping, and

---

[5] The usage of 'his', instead of 'her', or 'his/her', is virtually forced by the historic circumstances: starting with the 19th century, and going backwards in time, mathematics and other sciences were de facto out of reach for women. There will be only one woman among the mathematicians we will consider.

typically, in such cases, mystic beliefs did not adversely affect the practice of mathematics.

In Western societies during the last two centuries or more, mystic beliefs were often labeled as vantages of the ages of superstition. Beliefs in the genuineness of mystic or paranormal events were typically deemed as being not worth attention by any genuine scientist, mathematicians included. Those who have unwisely and openly declared their mystic affiliations have often been exposed at least to scorn. Hence, there existed and still exist ample reasons to keep mystic beliefs out of sight. In the case of Gödel, that meant almost total secretiveness until literally the day he thought he would die. In the cases of Brouwer and Newton, we had clear Dr. Jekyll and Mr. Hyde situations: they were mathematicians by day and practicing mystics by night, so to speak.

The effect of mysticism on math in Western societies follows along the above lines. Instant converts to mysticism tend to abruptly change their mathematical work. Pascal is one such example: his case is preserved by the sheer set of circumstances. Gödel's life seems to follow the same pattern; however, in his case, we do not know exactly what led him in that direction. We also have the case of Augustus De Morgan, whose mystical-mathematical life resembles Gödel's.

Mainstream Western society before the 18th century had a negative attitude towards mysticism, but the reasons were different compared to the contemptuous dismissal of mystic-related phenomena during more recent times. Mysticism was then often regarded as dangerous heresy, and the Inquisitorial punishment was far more severe. So, it was risky to one's well-being to declare belief in mystic doctrines, and it was even riskier to engage in any practice that was overtly mystic. Hence, our heroes of those times mostly kept their mystic beliefs to themselves or camouflaged them with conventional proclamations. Nevertheless, the influence of their mysticism on their mathematics is discernible. In the case of Kepler, this is obvious: the ideas for his discoveries are clearly linked to his mystic belief of the existence of celestial divine geometry, and astrology. In the case of Leibniz, this seems also to be true, though it is not so apparent: his main mathematical discoveries were linked to the concept of infinitesimal, the same one that appears in his ostensibly mystic scheme called monadology. The ideas of infinite and infinitely small still had at that time a mark of mysticism in their origin. In that respect, a similar analysis applies to Newton.

Ultimately, the problem of the linkage between mathematics and mysticism is a type of chicken-and-egg chronology puzzle. I am inclined to believe that mathematics and mysticism come as a package in most of the individual cases: you need to be a bit of a mystic to do math, and you need to know a bit of math to understand and take seriously the mystic tenets.

CHAPTER 3

Alexander Grothendieck

The Man Who Searched

The dream is the oldest oracle.

PLUTARCH

3.1 GROTHENDIECK: HIS EARLY LIFE, AND HIS PARENTS

Alexander Grothendieck died on November 13, 2014. He was 86. He had devoted the last 23 years of his life to 'harmonizing the ways of his mind and his heart'. We do not know if he succeeded. He certainly devoted more time to that than Prince Lubevetzky.

Alexander Grothendieck was born in Berlin on March 28, 1928, as Alexander Raddatz. Alf Raddatz was his mother's first husband. His mother, Hanka Grothendieck, who played a major role in his life, was the source of his last name later on. His father's last name was also not fixed from his birth; he was born as Alexander-Sasha Shapiro, and when he was exiled from Russia to Germany, he changed his name to Alexander-Sasha Tanaroff. The family, extended by Hanka's daughter from her first marriage, lived together during the first six years of junior Alexander's life. In 1934, at the time of the rise of the Nazis, the father had to flee and went to Paris. Hanka followed him, and the young Alexander lived as a foster child for four years during the war.

DOI: 10.1201/9781003282198-3

Alexander Grothendieck,
around the age of 5 years.

Sasha Tanaroff was above all a revolutionary. As a teenager, he participated in diversions against the Russian Tsarist regime. When his anarchist
group was caught, all of the members were executed, except himself; Sasha
was led to the execution wall every day for three weeks, and his life was
eventually spared due to his youth; he spent the next ten years in prison.
After the October Revolution, he found himself in a peasant army, fight-
ing against the Bolsheviks. Again he was caught and sentenced to death.
At about that time, he was wounded and lost his left arm. He managed to
escape and then flee to Germany and, as we mentioned, further to France.
Together with Hanka, they participated in the Spanish Revolution on the
side of the Republicans. After the end of the Spanish Civil War, he was
interned in Camp Vernet, a reception camp for fleeing Republicans at the
foot of the French Pyrenees. In 1942, he was extradited to the Germans,
then transported to Auschwitz, where his life ended.

Grothendieck's
parents

During his childhood, Alexander Grothendieck was mostly under the guidance of his mother. However, his father's rebellious lifestyle left a clear stamp on his own life. Sasha Tanaroff's bold and uncompromising life-changing moves were echoed in Alexander's wild quest to know himself, as he himself put it.

As mentioned earlier, Hanka Grothendieck exerted a major influence on her son. It was she who stayed with the young Alexander throughout most of his difficult childhood and teenage years. Both Hanka and Sasha had literary ambitions and talents. Hanka wrote a well-received autobiographical novel *Eine Frau* (*One Woman*). During various stages of his life, Alexander also engaged himself in literary pursuits, and the legacy of the talents of his parents is visible.

The relationship between Hanka and Alexander, as it seems, had deep undertones bordering on an Oedipus complex. Alexander wrote in his memoirs:[1]

> The second passion in my life was the search for woman. This passion often presented itself to me under the guise of a search for a companion. I could not distinguish the one from the other until around the time when it ceased, when I knew that that which I sought could not be found anywhere, or also: that I carried it within myself. My passion for woman could not really take wings until the death of my mother [in 1957].

The first passion was, of course, mathematics. Grothendieck was a singularly talented mathematician. After studying and self-studying mathematics as a teenager and as a young adult, and after rediscovering the Lebesgue integral along the way, he ended up in Paris in 1948. There he met some prominent mathematicians, including Henri Cartan, André Weil, and Laurent Schwartz. A couple of years later, Schwartz, fresh from receiving the Fields Medal, gave him a paper, jointly written by him and Jean Dieudonné and containing 14 open problems. It took Grothendieck a few months to solve all of them. All of them! Is it possible to be more brilliant?

The irony is that Grothendieck was not very interested in solving singular problems. He wanted to devise theories allowing him to gain a wide view of the whole field. And theories he found.

---

[1] [Sch], p. 3 (Chapter 2).

3.2 FROM A MATHEMATICIAN TO A MYSTIC

Grothendieck is credited with being the person who created modern algebraic geometry. For that and for his numerous other contributions, he was awarded the Fields Medal.

Alexander Grothendieck
(1970)

Two years after he received the Fields Medal, in 1968, a student revolution broke out in Paris. This sharply shifted Grothendieck's interests, and he started to be increasingly involved in social as well as ecological movements of that time. His mathematical interests started to slowly fade away. Between 1972 and 1973, he lived in a 'new age' commune, then moved to the small village of Villecun near Lodève, southern France, where he settled for the next six years (1973–1980). During this period he was still involved with mathematics, teaching math courses twice a week at the university in Montpellier. During the 1980s he continued to live in small villages, not yet completely isolated from social life.

While in Villecun, Grothendieck already leaned heavily in the direction of mysticism. Buddhist monks from the nearby rural commune of Olmet sometimes visited him. It was during this period that Grothendieck, as he put it, 'discovered meditation'. What this meant to him is not perfectly clear. He wrote the following enigmatic item in his list of the 'most important spiritual events in his life':[2]

15/16 Nov. 1976: collapse of the Image, discovery of meditation

---

[2] [Sch], Chapter 11.

It seems that he did not refer here to meditation as a basic consciousness-altering technique, for, as hinted above, since 1974 or earlier he was familiar with Buddhism, specifically with the Nipponzan Myohoji movement.

The reference to collapsing images, as cryptic-looking as it is, offers more clues about what he might have meant. The concept of 'collapsing of the image', or 'breaking of the image', has been mentioned many times in metaphysical or mystical sources. For example, the following quote is attributed to the founder of the (British) Society of Psychical Research, Frederic W. H. Myers:[3]

> freedom can only come through the deliberate process called 'The Breaking of the Image.'

From this and other similar statements by Myers, it is easy to deduce that the phrase 'the breaking of the image' refers to advancing to another, higher stage of existence. The reader may be inclined to dismiss Myers' statement as too occultist. If so, hold on: there will be ample opportunities to dismiss much more. Indeed, the context in which Myers is quoted is far more outrageous than the quote itself. We will defer further explanation[4] till later; first, we need to get used to 'miracles'.

Occultist or not, the virtually coincidental terminology used by Myers and Grothendieck is intriguing. In the context of Grothendieck's undeniable shift to mysticism, interpreting both phrases as indicating a move to a perceived higher state of consciousness is reasonable.

The notion of 'image' in this setting may then be interpreted as referring to the perceived external reality. The 'image' is a false mental picture, or *Maya*, according to the old Vedic texts. The 6th-century Buddhist monk *Bhavaviveka* (or *Bhaviveka*) expressed this in the following verse:[5]

> The form is like the foam,
> [],
> and the perceptions are like illusions (maya),
> this is declared by the truth-seer.

---

[3] Reference will be given in Section 6.4.
[4] Section 6.4.
[5] [Iid], p. 144.

The process of breaking or collapsing the Maya images is often mentioned as one of the first steps in the direction of achieving a mystic state of consciousness.

If the above interpretation of Grothendieck's short note is correct, then it seems plausible that he attained some kind of a deeper mystic state in November of 1976. Since this note is chronologically the first 'important spiritual event' that he mentions in his memoirs, his experience may have been strong enough to set him firmly along the mystic road. This may only be a conjecture; however, it is indisputable that he eventually went that way, determined to get as far as he could.

3.3 THE DREAMER

The 12th-century German mystic Henry Suso stated that in order to perfect one's life one should 'keep thyself pure from all images introduced from outside.'[6] In dreams keeping oneself 'pure from all external images' happens by default. Since dreams had special importance to Grothendieck, we stay with this subject for two more paragraphs and consider it from the mystic perspective.

The ancient philosopher Democritus wrote in his aptly titled treatise *On Images* (a fragment of which was preserved by Plutarch) that in dreams we are able to penetrate beyond the mundane images that are 'representations of the [] thoughts [] of the persons who originated them.' Again we see the mystic formula for spiritual progress in the direction of a union with the Ultimate Source: one must access the realm beyond thoughts and beyond their byproducts, images. Images are but provisional clues through which one infers, rather than perceives, what we call reality.

Dreams have a special place in mystic lore, where the claim that 'experience in dreams is more related to "reality" than what we see while awoken'[7] is taken seriously. A more conservative (mystic) claim is that dreams are more than merely random fantasies produced by the dreamer: they are a tangible link with one's inner consciousness.

Grothendieck considered dreams to be windows to knowing one's true self. It appears that his discovery of the spiritual potential of dreams

---

[6] [Sus], p. 135.

[7] [Robl], p. 481. Don Juan in *Journey to Ixtlan* by Carlos Castaneda concurs: '[Dreams] are more real [than waking life]', ([Cas], p. 92). I am, of course, aware of the allegations that Castaneda's books about Don Juan are fiction, not an anthropological investigation. I disagree with that claim as pertaining to his first two books.

happened in 1976: in a one-page note listing his main spiritual events, Grothendieck wrote:

18 Nov. 1976: reunion with my soul, entry of the dreamer[8]

True to this announcement, Grothendieck-the-dreamer started to diligently record and analyze his dreams. One example is the following dream, noted in Grothendieck's thousand-page-long autobiographical manuscript titled *Récoltes et Semailles* (*Gleanings and Sowings*, 1986). The dream, which happened in July 1984, made a strong impression on him:

> The dream I'm referring to had no scenario, no specific acts or activities. It contained but a single frozen image, one that was at the same time remarkably alive. It was a human head seen in profile, scanned from left to right. The head was of a mature man, beardless, with wild head wrapped around its brow like a bright powerful halo. The strongest impression made by this head was of a joyous, youthful vitality, which seemed to spring directly from the supple and vigorous arching of its neck (sensed more than seen). The facial expression was more that of a mischievous delinquent than of a responsible or settled adult, thrilled by the recollection of some trick he'd gotten away with or was about to do. It gave off an intense love of life, playful, content with itself.
>
> [] The perception of this head, or let us say of the atmosphere which it evoked, was extremely intense. [] There was only this intensely vital object, the man's head, and an awareness of that vitality.

When reading this note one gets the impression that it is a record of an apparition or a hallucination. In any case, however one interprets the event, it is evident that dreams were much more than subconscious fantasies to Grothendieck. This attitude, if nothing, is firmly mystic.

Grothendieck's view that dreams are much more than randomized fantasies is shared not only by mystics but also by some prominent scientists. For example, according to quantum physicist Fred Alan Wolf, dreams are sometimes visits of our consciousness to parallel realities, and, as such, they are windows to deep reality. Mystics take this claim one step further:

---

[8] [Sch], Chapter 11.

according to some of them the waking experience is an objectivized dream. Going even further in that direction, we find the following very intriguing statement:[9] 'Your daily experiences are the dreams that [your nightly, dreaming self] dreams.'

This last claim is not gratuitous! There is a huge library of recorded pre-cognitive or premonitive dreams: people dreaming of events that come to pass in the future that we experience. So, at least sometimes, our waking reality realizes our dreams.

The dream state itself has been the subject of many investigations. For example, there are scientific experiments statistically confirming that dreams enhance our ability to perceive at the subconscious level, so that, in such a way, they become a mode of extra-sensory perception (commonly acronymized to ESP).[10]

We mention that Grothendieck's philosophical views on human society, his place in it, and its future, were rather eccentric. He was convinced that a cataclysmic event would cleanse the planet by the end of the 20th century, clearing the terrain for an imminent golden age of humanity. According to Grothendieck, those who shared this prophecy were human mutants who were the precursors to the new age.

In 1991, Grothendieck withdrew completely from public and social life. His withdrawal was so definite that very few people even knew where he lived during the remaining 23 years of his life.

He made a short intervention in 2010, requesting a complete halt on any publishing of his work and asking for a complete withdrawal of his already published writings from libraries. In this respect, it seems, he followed the doctrines of enlightened Buddhism, according to which in order to achieve the ultimate spiritual freedom, one should sever all earthly ties, especially those that may be interpreted as being connected to pride. Pride is a limiting vice according to Buddhism, with which Grothendieck was certainly familiar. Many other mystic teachings concur; the following segment of the poem 'Wake up' written about 800 years ago by the great Sufi poet Rumi, may be a summary of the motivation behind Grothendieck's seemingly irrational actions by the end of his life:

It's a blessing of the ego to be humbled,
Alas for the one who rose into a mountain of arrogance,
Know that pride is a killing poison.

---

[9] [Rob2], p. 135.
[10] See, for example, [Ulm].

The influence of his shift to mysticism on Grothendieck's mathematical life is evident: his mathematical research seems to have stopped entirely. He did not publish anything during the last 23 years of his life. He acted as if he had changed not only his life but also his personality. We do not know if he was still engaged in any kind of mathematics during this period. In this sense, his life resembles Pascal's, except that in the case of the latter, the associated events, indeed Pascal's entire life, resembled an accelerated version of Grothendieck's life story.

Were the 'spiritual events' that Grothendieck recorded in his diaries mystic experiences? Did he travel all the way along the mystic paths? Even though there is no doubt that he was a mystic according to our definition, there is little that he made available as clues to assess the degree of his mysticism. It seems that his hermit-life existence during the last many years of his life was a deliberate attempt to self-initiate himself into the art of true mysticism. However, there is no hard information available to support this claim. So, Alexander Grothendieck, the extraordinary person that he was, will be set aside as a temporarily unsolvable enigma.

CHAPTER 4

Kurt Gödel

The Man Who Proved

There are spirits which have no body but can communicate and influence the world.

KURT GÖDEL

4.1 FROM GROTHENDIECK TO GÖDEL

According to Alexander Grothendieck, 'each person has a "mission" and that an important part of this mission consists of finding one's own self.'[1] Kurt Gödel's corresponding statement was even more sweeping; he believed that 'if you know yourself you know everything.'[2] The philosophical views of these two great people suggest that they did not refer to mundane self-psychoanalysis in their references to self-knowledge but rather to the knowledge of oneself in the context of the greater spiritual realm.

The spiritual meaning of 'knowing oneself' is a very old theme. Origen, the 2nd–3rd-century Christian thinker and mystic, considered self-knowledge as the first stage (out of nine) of attaining the ultimate mystical union. He was following the wisdom of Jesus Christ (Gospel of Thomas): 'And the Kingdom of Heaven is within you; and whomsoever shall know himself shall find it.' Pythagoras, the ultimate mystic among

---

[1] [Sch]*.
[2] [Wan], p. 150.

DOI: 10.1201/9781003282198-4

mathematicians, allegedly said virtually the same:[3] 'Man know thyself; then thou shalt know the Universe and God.' Going further back in time, the inscription above the temple of the Oracle of Delphi reads, 'Know thyself'. The priestess who was the Oracle of Delphi was an epitome of a true mystic: so 'know thyself' can only be interpreted from that perspective. Even more explicit is the ancient (15th–10th century BC) Vedas' teaching that, 'the Absolute Reality [] dwells fully in the human soul, and that this reality is meant to be known.'[4]

None of the above refers to the psychological self, or the ego; from the mystic vantage point, Self extends beyond the mundane realm of senses. Knowing 'thyself' is, in essence, equivalent to accessing the ultimate Self, or the Supreme Consciousness.

Both Grothendieck and Gödel experienced a kind of midlife philosophical crisis, at the end of which they both departed in the direction of mysticism. This is a pattern that we will encounter many times. The facile explanation that with age, and when facing the nearer and nearer prospect of impending death, one searches for a philosophy of hope, is not satisfactory. None of the mystics-mathematicians we will encounter, and no late-blooming mystics in general, was obsessed with the thought of annihilation. Gödel's mysticism, in particular, was most certainly not caused by thanatophobia. He had a much simpler explanation: he said that with age, people think better. The obvious fact is that, after 30, 40, 50 years of experiencing life, it is only natural to try to distance oneself from the inertia of living and to look into ontological matters, such as the meaning of life. This is especially true if one experiences an event that could be interpreted as being of mystic origin. Materialism does not offer satisfactory answers in many such cases.

Grothendieck and Gödel arrived at their partially overlapping mystical beliefs from distant points of departure: their personalities and their lives were rather incongruent. It seems from what we know that Gödel was an introvert who maintained very few social contacts. Grothendieck, on the other hand, spent a few years of his life in a New Age community, where free sex was a norm. Gödel followed the dress codes of his time and in many photos appears with a formal tie; Grothendieck was the first ever to give a presentation at the Romanian Academy of Sciences dressed in shorts; and where Gödel avoided publicly revealing his work on some ostensibly controversial subjects until literally the day he thought

---

[3] [Str], p. 282; no source given.
[4] [Star], p. 3.

he would die, Grothendieck, while he was socially active, almost casually created controversies along his way.

Grothendieck's outward life was a sequence of discontinuous (but converging) forays. By contrast, Gödel's outer life appears to have been a fairly linear road leading from mathematics to metaphysical philosophy. In Gödel's outer life, there was no social flamboyancy that was Grothendieck's trademark. However, as it seems, Gödel's intimate life hides deep secrets that are only hinted at by the rather scarce information that is available.

Subjectively, Gödel means more to me. I consider his incompleteness theorem – sufficiently large consistent theories are always incomplete – to absolutely be the most important (precise) epistemological result of our civilization. Moreover, his metaphysical philosophy and his mysticism – to the extent that we know them – are more relevant to our project than Grothendieck's. We will use Gödel's views as a bridge from modern times to the less constrained social paradigms of the old days. So, we will devote more time to him.

4.2 A SHORT BIOGRAPHY OF GÖDEL

When Gödel was born (1906), he automatically became a citizen of the Austro-Hungarian Empire. His parents were relatively affluent, and Gödel had a sheltered childhood, receiving a solid education at the German Lutheran school in Brün (Brno). He got the highest marks in all subjects during his first 12 years of education, except in gymnastics and, as often happens with bright children at the mercy of mediocre teachers, once in mathematics.

Kurt Gödel, 1930?

When he was 12, the Austro-Hungarian Empire dissolved, and he ended up a Czechoslovakian citizen. Six years later he was an Austrian living and studying in Vienna. During the six preceding years, his family cultivated their Austrian-German heritage, and Gödel later said that he felt like a refugee in the predominantly Czech-speaking country.

Gödel enrolled in the University of Vienna in the fall of 1924. It took him five years to finish his studies, all the way to the defense of his doctoral dissertation, in which he proved the completeness of predicate logic. Two years later, in 1931, he published the proofs of his first incompleteness theorem (sufficiently large consistent theories cannot include all true statements in the language of that theory[5]) and, in a separate article, the proof of the second incompleteness theorem (certain theories that include the statement of their consistency must be inconsistent). This was the peak of his mathematical career. He was only 25 years old at that time.

His academic tenure started in 1933 when he became a Privatdozent.[6] He published extensively during these years, discovering and proving other important theorems along the way. Then there follows a conspicuous gap of two to three years, a period of time during which Gödel did not submit any research articles for publication; the two articles that appeared in 1936 were originally presented two years earlier. We will come back to discuss and interpret this anomaly later on.

His work was apparently restarted in 1938 when he published a few important papers. In September of 1939, he married Adele Porkert, after a long relationship extending some 11 years. Adele had virtually no influence on Gödel's intellectual life. However, as we will see later, she had a huge influence on Gödel's life in general.

In 1938, Austria was annexed by Germany, and Gödel was transformed overnight into a citizen of Germany. Four years later he moved to America (via Siberia, Mongolia, and China) accepting a position at the Institute for Advanced Study in Princeton. He settled in Princeton for the rest of his life, never visiting Europe again, and rarely traveling at all. He died in 1978 as an American citizen, the fifth citizenship in his life.

---

[5] In layman's terms, there never will be the book of all truths, 'on earth or in heavens'.
[6] A junior professorial position.

4.3 GÖDEL'S PRINCETON TRANSFORMATION AND THE IMPACT OF MYSTICISM ON HIS MATHEMATICS

When Gödel moved to America at the age of 36, all of his most important mathematical achievements were behind him. At Princeton, he found stability in his life, and his academic position was secure. It appears that these were the conditions necessary for the sharp change in his manifested interests. Our not-too-speculative thesis is that there was an earlier event that precipitated his metamorphosis from a first-class research mathematician to a philosopher studying metaphysics, occultism, and mysticism.

The first hints of a major change in his intellectual life can be found in the articles that he published while at Princeton. There were only two published articles between 1942 and 1947: the first one (titled *Russell's Mathematical Logic*, 1944) was an expository report on the foundation of mathematics based on the correspondence between Gödel and the mathematician Arend Heyting that had happened more than a decade earlier; the second (*What is Cantor's continuum problem?* 1947) is the only popular exposition written by Gödel. Between 1948 and 1950, Gödel published three papers on general relativity theory. This in itself indicates a major shift in his intellectual priorities: he was now concerned with the problem of defining the basic properties of perceived reality. Moreover, as his biographer Wang and others have noted, Gödel's work in physics was inspired by Immanuel Kant's philosophy of space and time, and Kant's interest lay mainly in metaphysics. Thus, through his physics papers, we can discern that Gödel's philosophical affinities already coalesced in the direction of metaphysics.

At about that time (1950), Gödel stopped publishing research papers altogether. His interests in philosophy, particularly in metaphysics, started to dominate his intellectual life.

Gödel was not a run-of-the-mill philosopher; his ingenuity and his preeminence in the field of logic gave him a unique perspective, and it is then not surprising at all that his metaphysical ideas and goals turned out to be unconventional as well. He wanted to put metaphysics on a firm theoretical base, starting with formal axioms. He said that he wanted to devise, 'an exact theory in philosophy that is comparable to Newton's in physics.'[7]

Gödel worked on this theory secretly; as we mentioned earlier, he did not like controversies, and he was afraid – likely rightfully so – that he would be misunderstood. He anticipated that people would take his work

---
[7] [Wan], p. 152.

as expressing belief in God, rather than what he intended it to be: a formal investigation in logic.

His ontological theory was made public in 1970 and only because Gödel thought that he was about to die. It is, indeed, an inception of a formal theory, starting with axioms that branch into various theorems.[8] The intended model was the transcendent, deeper reality. God was a primitive notion in this theory – in the same sense as 'point' is a primitive notion in axiomatic geometry. From his axioms, it follows that there exists a God-like entity. This is often interpreted as Gödel's proof of God's existence – a misinterpretation that in a way justifies Gödel's reluctance to publicize his theory. He did not prove it, and he would have been the last person not to know what a formal theory meant. He merely devised a sound theory having in mind a model and an interpretation of the notion of God, where God's existence would be a consequence of the fact that the axioms would be true in that model.

Ax. 1. $\{P(\varphi) \wedge \Box \forall x[\varphi(x) \to \psi(x)]\} \to P(\psi)$
Ax. 2. $P(\neg\varphi) \leftrightarrow \neg P(\varphi)$
Th. 1. $P(\varphi) \to \Diamond \exists x[\varphi(x)]$
Df. 1. $G(x) \iff \forall\varphi[P(\varphi) \to \varphi(x)]$
Ax. 3. $P(G)$
Th. 2. $\Diamond \exists x\, G(x)$
Df. 2. $\varphi \text{ ess } x \iff \varphi(x) \wedge \forall\psi\{\psi(x) \to \Box \forall y[\varphi(y) \to \psi(y)]\}$
Ax. 4. $P(\varphi) \to \Box P(\varphi)$
Th. 3. $G(x) \to G \text{ ess } x$
Df. 3. $E(x) \iff \forall\varphi[\varphi \text{ ess } x \to \Box \exists y\, \varphi(y)]$
Ax. 5. $P(E)$
Th. 4. $\Box \exists x\, G(x)$

Gödel's ontological theory. Th. 4 can be interpreted as follows: It is a provable statement that there exists a God-like entity.

However one interprets his ontological theory, one thing is certain: Gödel's mathematical interests during the last two decades of his life were only a byproduct and a function of his passion for metaphysical philosophy. If we accept that his interest in metaphysics was motivated by his mysticism, which is what we will conjecture in the next section, then the ontological theory can be viewed as an idealized and formalized rational mysticism.

Gödel's interest in metaphysics was most certainly not brought about by his religiousness. He was not religious in the common sense of the word. Even though he was a nominal Lutheran, he was not a member of any congregation, and he did not go to church. He wrote in his papers[9] – in Gabelsberger shorthand that was deciphered after his death – that, 'religions are, for the most part, bad – but religion is not'. For Gödel, *religions*

[8] It is now easily accessible over the web.
[9] [Wan2], p. 316.

were associated with dogmas, while one can surmise from his writing that the main attribute of his definition of *religion* is the belief in the existence of God. His ontological theories strongly suggest that Gödel himself had such a belief. In fact, more is true: a few times he declared explicitly that he was a theist. Believing in the existence of an omnipresent and originating Supreme Being, or God, does not imply that the believer is religious.

Some people based their claim that Gödel was religious on a statement made by his wife after his death. She said that Gödel used to read the Bible every Sunday morning, presumably by the end of his life. His wife, Adele, did not share Gödel's mystic interests, and she sometimes joked publicly about them. They were very different personalities, both intellectually and emotionally. With her claim, the truthfulness of which we do not doubt, she may have inadvertently provided pretense to those who were all too ready to misinterpret that part of Gödel's life as irrational and to amputate it from Gödel-the-scientist. Reading the Bible every Sunday morning does not make one a religious follower. We will come back to this topic in the next section.

Gödel's flair for unorthodox formal theories did not start with his ontological scheme. The following anecdote shows that he was deeply involved with metaphysical axiomatizing for at least a quarter of a century. During the 1950s, Gödel developed a strong friendship with a man who was a generation older than him. They would meet almost every day for many years, walk together, and chat amiably about various subjects of common interest.

His friend was Albert Einstein. At that time, 1950–1953, Einstein had an assistant, the mathematician Ernst Gabor Strauss. Strauss tells us of yet another of Gödel's axioms, the *no-accident* axiom:[10]

> Gödel has an interesting axiom by which he looked at the world; namely, that nothing that happens in it is due to accident (or stupidity). If you really take that axiom seriously all the strange theories that Gödel believed in become absolutely necessary. I tried several times to challenge him, but there was no way out. I mean, from Gödel axioms they all followed.

---

[10] From *Some Strangeness in the Proportion; A Centennial Symposium to Celebrate the Achievements of Albert Einstein*, edited by Harry Wolf, p. 485.

Strauss did not elaborate on Gödel's 'strange theories'. A hint is given in Wang's books,[11] where he quotes Gödel himself:

> the world and everything in it has meaning and reason, and in particular a good and indubitable meaning. It follows immediately that our worldly existence, since it has in itself at most a very dubious meaning, can only be means to the end of another existence.

This quote actually starts with 'The theological worldview is the idea that …' I have omitted it not in order to slant the account but because it follows from Strauss's story and from the next section that Gödel actually believed in the modified declaration. It was his style to depersonalize his statements in order to make them less controversial.

Gödel described himself as a Neoplatonist, and Neoplatonism started with the 3rd-century Christian mystic and philosopher Plotinus. Hence, it is not surprising that Plotinus adhered to a similar no-accident axiom:[12] 'To attribute the being [] of [the universe] to accident [] is unreasonable and belongs to a man without intelligence and perception.' This standpoint, of course, is in the background of many theological teachings too.

It is clear that Gödel's no-accident axiom leads to a complete revision of the materialistic paradigms. To mention a single example, our own existence becomes predestined, implying in turn the existence of God-like creators of reality. Einstein once proclaimed aphoristically that, 'God does not play dice.' So, one would expect that he would be on the same page as Gödel on this issue. However, as it seems from Strauss' account, Gödel's axiom and its consequences were beyond Einstein's tolerance. The last time Strauss met Einstein, the latter said:[13] 'You know, Gödel has gone completely crazy.'

All in all, it is certain that something happened in Gödel's life that profoundly affected his mathematical and philosophical interests, and given the scope and direction of the change of his interests, it seems likely that there was some kind of influential mystic event in his life. If we do not postulate this hypothesis then there are irreconcilable incongruences in Gödel's life trajectory. First, one does not go far in the direction of

---

[11] [Wan], p. 216–217.
[12] [Plo], Volume 3, p. 43.
[13] [Wan], p. 32.

mysticism, or in the direction of the occult, without a powerful motive and at the expense of virtually all of the past professional mathematical life. Math is rather addictive to let go of so casually. Second – and we will elaborate in the next section – Gödel's views on reality in general went far beyond merely rejecting materialism in order to be explainable in purely intellectual terms.

4.4 GÖDEL, MYSTICISM, OCCULT

Plato and Plotinus were two old thinkers who significantly impacted Gödel's philosophy. However, according to Gödel himself, it was Leibniz who influenced him the most. We will talk much more about Leibniz in Chapter 10. Here we mention an episode related to the fourth philosopher who impacted Gödel's views: Edmund Husserl (1859–1938), the founder of the school of phenomenology (the study of appearances). His main idea is the postulate that certain phenomena, hallucinations included, should be studied in their own terms, independent from the problem of their origin. This cautious thesis resonated with Gödel. However, we mention Husserl here because of a different reason. Apparently, between the years 1906 and 1910, after a period of psychological crisis, there happened a sudden breakthrough in Husserl's work. Gödel interpreted the change as being a result of a 'sudden illumination', and mentioned Descartes as another example of a philosopher who experienced a similar episode. It is not easy from the available data to give a categorical interpretation of what Gödel meant by 'sudden illumination', though the comparison with Descartes, whom we will discuss in Chapter 12, suggests that he may have been referring to some kind of a mystical experience. Gödel's biographer, Hao Wang, does not consider this explanation at all. He speculates:[14] 'Perhaps Husserl was able, after persistent efforts over many years, to understand what is truly fundamental in philosophical investigations.' The philosopher Richard Tieszen's view is similar: according to him, Husserl simply got a new idea by studying philosophy. We mention this episode in order to illustrate the prevailing bias in conventional interpretations of events or claims with mystic connotations. Both Wang and Tieszen play it safe and present 'normal' explanations, implying that Gödel was referring to a kind of eureka moment. However, this was not true: Gödel stated[15] with regret that he himself had 'never had such an experience'. Gödel, the man

---

[14] [Wan2], p. 293.
[15] [Wan2], p. 169.

who discovered the incompleteness theorem, could not have meant that he himself had not experienced a 'eureka moment'.

Tieszen wrote the book *After Gödel*, where he discusses Gödel's philosophy. In the introduction of his book, he writes:[16]

> Many of Gödel's comments on religion, immortality of the soul, and similar matters have been widely quoted. There are also comments and certain activities of Gödel that are just paranoid and bizarre. Readers who are interested in this more 'scandalous' material will have to look elsewhere.

Tieszen was embarrassed for Gödel, and he sanitized his philosophy. This is typical for biographers of notable scientists: anything spiritual or mystic is equated with superstition, it is deemed antithetical to the modern scientific paradigm, not fitting the stature of the scientist in question, and is thereby unceremoniously suppressed, distorted, or eliminated. We will encounter a few other such cases later on.

In any case, I followed Tieszen's advice and looked elsewhere. There is indeed, and fortunately, a substantial quantity of 'scandalous material' about Gödel. I am convinced that these sources matter if one wants to understand this great man and his views. Cramming Gödel's philosophy within the narrow frames of a paradigm that excludes mysticism, the concept of immortality, and other 'scandalous material', is bound to distort it.

As we mentioned earlier, Hao Wang, who was not only Gödel's biographer but also his friend, was a skeptic as well. However, to his credit, and thankfully so, he duly reported Gödel's 'scandalous' claims. His two books, *Reflections on Kurt Gödel* and *A Logical Journey; From Gödel to Philosophy*, are our main sources of Gödel's mystic or 'occultish' claims given in this section.

Not surprisingly, materialism was not to Gödel's liking. In his personal notes, he stated[17] emphatically: 'Materialism is false.' According to him, the first philosophy is metaphysics, centered around the notions of God and soul. For Gödel, the existence of God implies as a corollary that the world's origin and structure are rational. His no-accident axiom is a consequence of this theory. There are other consequences that he mentioned

---

[16] [Tie], p. x.
[17] [Wan2], p. 316.

and with which he resolutely entered through the doors of rational mysticism. Here is an insert[18] from a letter from Gödel to his mother Marianne:

> In your last letter you pose the weighty question whether I believe we shall see each other again [in the hereafter]. [] If the world is rationally organized and has a sense, then that must be so. [] But do we have reason to assume that the world is rationally organized? I think so.

Thus, Gödel indeed believed in the immortality of the soul, or of the mind, as opposed to the ephemerality of the body. Gödel told[19] Wang that he thought of mind as 'of one person living forever'. In this context 'person' is, of course, separate from matter; in 1972 he stated[20] this explicitly: 'Mind is separate from matter: it is a separate object.'

Immortality of the soul implies its post-mortem and, indirectly, pre-natal existence. Hence, argues Gödel, there is a domain of existence of individual spirits of the souls, as well as of 'collective minds'. Karl Jung calls a similar concept 'collective consciousness'. Gödel talks about this directly; he wrote in his Gabelsberger shorthand notes:[21] 'There are other worlds and rational beings of a different and higher kind' and, further, that 'the world in which we live is not the only one in which we shall live or have lived'.

It is clear that by the end of his life, Gödel dropped his rationalistic guard and started to open up. Wang notes the following statement by Gödel, made less than a year before he died:[22] 'There are spirits which have no body but can communicate and influence the world. They keep [themselves] in the background today and are not known. It was different in antiquity and in the Middle Ages, when there were miracles.' Here Gödel comes across as much more than a mere rational mystic: his theory reaches all the way to what today is called occultism. Note the categorical claim the asserting existence of body-less spirits. Gödel was too diligent to allow himself to state such an unequivocal existential statement based only on his reading, for in that case it would have been much closer to his character to condition the claim with 'I believe' or with 'it is reasonable

---

[18] [Göd], p. 429–431.
[19] [Wan], p. 150.
[20] [Wan2], p. 191.
[21] Ibid, p. 316.
[22] Ibid, p. 152.

to propose'. His views only on this subject support the thesis that some ostensibly mystical event happened in his life. One does not come to the conclusion that body-less spirits exist and manifest in our world by reading ghost stories.

We point out again that in 1976 Gödel was most certainly aware that his days on earth were numbered. In light of his previous episode of near death, when he decided to reveal his ontological proof, it is likely that he felt he had nothing to lose by being more direct and forthcoming with his personal views.

Also note Gödel's reference to miracles. He chose his words carefully; he once stated that 'everything goes only by probability'.[23] Yet, he did not qualify his claim about miracles in any way but rather flatly postulated their existence. That makes him at least a rational mystic galore.

Gödel implied, indeed he said outright multiple times, that he believed the soul, or the mind, survives physical death. With this in mind, we revisit his Bible-reading episodes. The 'miracles' aside, it is relatively safe to postulate that some of the saintly beings featured in the Bible existed. If so, then according to Gödel himself, their spiritual essence continued to exist. Hence his reading of the Bible may be interpreted not as a ritual act of a religious follower but rather as a way to interact with real beings that he himself considered as having the capacity to connect with the material world. Given Gödel's profound mystic disposition, this interpretation may not be far off the mark. Moreover, in view of his belief in the genuineness of miracles, it is likely that he considered the stories told in the old Christian scriptures as accounts of paranormal events, rather than as allegories or fiction. He might have agreed with Isaac Newton, who believed that the Bible was a source of ancient wisdom.

4.5 WAS GÖDEL A MYSTIC?

Was Gödel only a rational mystic, or did he travel along the mystic roads? There is circumstantial evidence that the latter might be true.

Almost all mystic philosophers have been propelled in that direction by some kind of inner experience. Many of them had good reasons to hush up anything of that sort, be it because of the inquisitorial retributions or because of contemporary stigma.

There was indeed a momentous personal event that happened in Gödel's life. Before we discuss it, we devote more attention to his almost lifelong

---

[23] [Wan2], p. 169.

companion Adele. She was of a humbler origin than him, and Gödel's mother was against their relationship. Moreover, Gödel's close relatives did not look indifferently upon the fact that Adele was seven years older than Kurt. Nevertheless, they eventually married. Adele knew things about Gödel that nobody else did. And on rare occasions, she did reveal a few interesting facts about Kurt's life.

As we noted earlier, Adele never shared Kurt's interests and was mostly an external observer of his intellectual pursuits. However, at times, she did not mind making light of his interests. Thus, according to Georg Kreisel,[24] on one occasion she made fun of Gödel reading about ghosts and demons. The event suggests much more than an interest in occult stories. Since this episode happened in the 1950s, long before Gödel went public with his mystical claims, it also suggests a certain chronology of events in Gödel's life. One does not read books about 'ghosts' and then becomes a mystic. To read books about such a banal subject, it is necessary to be a mystic of some sort, to begin with. Being interested in 'ghosts' could then be considered as an attempt to understand some aspects of the mystic experience. There is little doubt that in the 1950s, Gödel was already a closet mystic.

Gödel valued Adele's role in his life. More than a hint in that direction is given in the following anecdote, narrated by his biographer Hao Wang;[25] it concerns Alan Turing (who, as we will see later, was also interested in paranormal events, to the extent that he felt sufficiently motivated to write a pamphlet about the subject).

> In September 1956 Georg Kreisel took me to their house for afternoon tea. Adele was present but did not say much. I remember that we discussed Turing's suicide and that Gödel asked whether Turing was married. On being told that he was not, he said, 'Perhaps he wanted to get married but could not.' This observation indicated to me the importance Gödel attributed to marriage for a man's [] life and death.

Wang was probably right in his observation. However, there is much more in Gödel's commentary on Turing's suicide than what meets the eye: at one point in time, Adele actually prevented Gödel from committing suicide. This dark episode of Gödel's life was revealed to Wang by Adele a

24 [Wan2], p. 49.
25 Ibid, p. 51.

few days after had Gödel died. Apparently at one point of time before they married, Gödel was taken to a mental institution, against his will. Adele told Wang that during one of her visits, Gödel was in such a desperate state that he attempted to jump from a high window, and she barely managed to prevent him from realizing his moribund intentions.

Gödel's wedding with
Adele, 1939

We do not know other details of the ordeal, and we do not know why Gödel was taken to the mental institution. It may have been a mental breakdown of some sort – though I personally doubt this explanation. It was very easy to end up in a mental institution in the 1930s. In fact, this did not much change over the following 40 years: in the 1970s, eight researchers led by David Rosenhan checked into eight different hospitals and declared that they had 'heard voices' and that that was their single medical concern. Seven of them were immediately diagnosed as schizophrenics, while the eighth was diagnosed with 'manic depressive psychosis'. On the basis of that single declaration, even though they behaved normally, they were hospitalized for up to two months.[26] Gödel was conspicuously quiet about the personal experiences that had led him to the mental institution. Had he burned himself by frankly sharing various episodes in his life with some 'well-intentioned people' who tried to help him regain his 'sanity'?

---

[26] David Rosenhan: *On Being Sane in Insane Places*, 1973.

The year 1936 is enigmatic in Gödel's life. As noted in Gödel's bibliography by Dawson,[27] a few things happened – or did not happen – that year. Up to 1936, Gödel was a regular reviewer for the German Mathematical Reviews (*Zentralblatt für Mathematik und ihre Grenzgebiete*); in 1936 this abruptly stopped. Moreover, it is visible from his bibliography that Gödel did not do any research work in 1936 and 1937.

This episode of him ending up in a mental hospital happened eight years into his relationship with Adele and three years before they married: 1936 is the 'right' timing for such a deeply personal and binding episode in their common life.

It does not follow that this is the timing of Gödel's hypothetical mystic event. However, when these subjective experiences occur, then they usually happen in critical times of the lives of the experiencers. Hence, 1936 or not long before is a good candidate for the timing of such a scenario in the case of Gödel.

Is it a far shot to conjecture that a proto-mystical phenomenon was experienced by Gödel in 1936 or earlier? Not at all! Recall the statistics mentioned in the preliminary chapters regarding so-called hallucinations. These events are fairly common, if not ubiquitous, and a large proportion of people experience them at least once in their lifetime. The problem, as we have noted earlier, is not in the singularity of these events but rather in the interpretation. In the case of persons who interpret them 'correctly', as in the case of Ivan in Chapter 1, nothing much happens. On the other hand, an 'incorrect' interpretation is a life-changer. And it takes much less than talking with Jesus for a couple of hours, as in the case of Pascal. Gödel's inner psyche was deeply nonconformist, hence the thesis that he would have chosen the 'wrong', non-materialistic interpretation as reasonable. In light of the large probability that he experienced a proto-mystic event or some kind of *hallucination*, and considering Gödel's statements regarding *spirits*, conjecturing the existence of a mystically interpreted event in his life is not as far fetched as it appears at first sight.

We postulate without proof that Gödel was more than a rational mystic; he was an experiencer of some sort. How far he traveled along mystic paths is not clear, but it is doubtful that he went very far.

27 [Daw]*

CHAPTER 5

Alan Turing and the 'Joke'

Once one has accepted [telepathy, clairvoyance, precognition, and psychokinesis] it does not seem a very big step to believe in ghosts …

ALAN TURING

5.1 TURING MACHINES

This is (a code for) a Turing machine...   ...and this is Alan Turing, 1951

DOI: 10.1201/9781003282198-5

Turing machines are the ancestors of modern computers. The hardware of a Turing machine consists of an infinite strip of squares, each containing either a 0 or 1 (or either of any two distinct symbols) and a scanner that can be in various states. The software or the 'code' of a Turing machine consists of a finite set of instructions that directs the scanner to act in a way that depends on its state and on the contents of the scanned square. In the example shown in the figure above, there are two states: state 1 (the starting state), indicated by a teardrop, and state 2, indicated by an upside-down teardrop. Shaded squares are assumed to contain 1, blank squares 0. The instructions are as follows (from left to right): when the scanner is in state 1 and when the scanned square contains 1, move the scanner one square to the right without changing the state; when the scanner is in state 1 and when the scanned square contains 0, change this 0 to 1, move the scanner one square to the right, and change the state to 2; when the scanner is in state 2 scanning 1 (third column in the diagram) or scanning 0 (last column in the diagram), do not do anything. We can now see that if this machine starts with exactly n-many consecutive squares containing 1 and scans the leftmost 1, then when it stops there will be $n + 1$ consecutive 1s. So, this Turing machine computes the function $f(n) = n+1$, for $n = 0,1,2,...$ and stops in state 2, scanning the rightmost 1.

The original purpose of Turing machines was purely mathematical: Turing needed them in order to precisely define the concept of computability and then to answer[1] in the negative the so-called *decision problem*, as stated by David Hilbert during a mathematical congress in 1928. Roughly speaking, the question was about the necessity of the existence of an algorithm that decides if a statement is true or false.

Note: The mathematics presented in our narrative is mainly for the purpose of illustrating the work of the respective mathematicians. It is chiefly independent from the rest of the narrative, and thereby it may be safely skipped without adversely affecting the readability of the rest of our story. This note applies throughout this book.

Having the concept of the Turing machine at hand, the main idea of Turing's proof of the non-existence of such an algorithm[2] is relatively simple, and we will outline it below. To that goal, consider the following Turing machine:

---

[1] The same result, but with different proof, was published by Alonzo Church in 1936.
[2] Published in 1937.

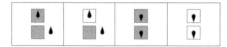

Comparing it with the first one, we notice a *slight* difference: this one doesn't change from state 1 to state 2 and, as a consequence, keeps printing 1s without stopping (the second state will not play any role here). So, some Turing machines stop, others do not stop.

Since every Turing machine is defined by a finite set of instructions, there are countably many of them. In other words, they can be enumerated by the numbers 1,2,3, …, and denoted $T_1, T_2, T_3,...$. Consider now the 'halting' function

$$f(n) = \begin{cases} 1 & \text{if } T_n \text{ stops when applied to a string of } n \text{ many 1s} \\ 0 & \text{if } T_n \text{ does not stop when applied to a string of } n \text{ many 1s} \end{cases}$$

Assume that there is a Turing machine T_f that computes this function. This means that if we apply the machine T_f to a string of n consecutive 1s, it will eventually produce a single 1 and stop, or it will produce a blank tape and stop. The former happens if T_n stops when applied to a string of n consecutive 1s, the latter if T_n does not stop.

Consider the function $g(n) = \begin{cases} 1 & \text{if } f(n) = 0 \\ 0 & \text{if } f(n) = 1 \end{cases}$. Since $f(n)$ is either 1 or 0, then when the machine T_f is applied to n consecutive 1s, it either stops in some previously unused state s_j while scanning the only 1 in the tape, or it stops in a previously unused state s_k while scanning any square in the blank tape. Now change the code for T_f as follows: when in the state s_j while scanning 1, change the state s_j to a new state s' which changes 1 to 0 and stops; when in the state s_k while scanning 0, change the state s_k to a new state s'' which changes 0 to 1 and stops. It is then obvious that this new machine computes the function g. So, we call it T_g. Hence, for every positive integer n, if $f(n) = 0$ (so that $g(n) = 1$), then T_g applied to n consecutive 1s produces a single 1 and stops, and if $f(n) = 1$ (so that $g(n) = 0$), then T_g applied to n consecutive 1s produces a blank tape and stops.

Here is, then, the climax of the proof: Since we have enumerated *all* Turing machines, T_g must be T_m for some m. According to the definition of the function f,

$$f(m) = \begin{cases} 1 & \text{if } T_m \text{ stops when applied to a string of } m \text{ many 1s} \\ 0 & \text{if } T_m \text{ does not stop when applied to a string of } m \text{ many 1s} \end{cases}$$

Now replace T_m by T_g in this equation and observe that what we get is clearly incompatible with the assumption that the Turing machine T_g computes the function g. So our hypothesis that T_f exists produces a contradiction. This means that there is no procedure to decide if the Turing machine #n halts or not when applied to a string of n-many consecutive 1s.

The self-referential strategy of the last part of Turing's proof is essentially the same as the argument used in Gödel's proof of his theorems. The two great men never met; Turing was a student at Princeton in 1936–1938, a few years before Gödel settled there.

Turing's ideas and his machine eventually gave rise to a slew of mathematical results, as was, for example, the unsolvability of the *word problem*[3] in group theory. In another direction, the machines and the associated codes evolved into modern computers.

5.2 A SHORT BIOGRAPHY

Alan Turing was born in London, England, on June 23, 1912. When he was one year old, his mother went to India to join his father, leaving Alan with friends and relatives of the family. Despite this, Alan, like Grothendieck and Gödel, had a rather strong connection with his mother, visiting her often and corresponding with her throughout his adult life.

His basic education was in a public school, where he was placed at the bottom group of students in subjects that did not interest him (for example, he did not care about learning Greek). In mathematics, however, he showed his talents early: when he was 15 years old, he rediscovered (without calculus) the series $tan^{-1}x = x - \dfrac{x^3}{3} + \dfrac{x^5}{5} - \dfrac{x^7}{7} + \cdots$

In 1931, he entered King's College where he studied mathematics. He graduated in 1934.

In 1936, he enrolled in graduate studies in Princeton, USA; his supervisor was Alonzo Church, the very person who was the first to establish the undecidability procedure (in arithmetic). He got his Ph.D. in 1938 and afterward returned to England.

---

[3] There is no algorithm that decides if a word on a set of generators of a group defines the trivial element

In 1939, at the onset of the Second World War, Turing started working as a leader of a code-breaking team. He was one of the principal designers of a machine called Bombe, a proto-computer of a sort, which played a pivotal role in deciphering German military/naval messages during the war.

Immediately after the war and as an acknowledgment of his contributions, he was awarded the title Officer of the Most Excellent Order of the British Empire. A few years later, in 1951, he was elected a Fellow of the Royal Society of London. Between the two honors he designed the first ever set of specifications for an electronic, stored-program, digital computer, and wrote the first ever programming manual.

Since 1951, he held a special position at the University of Manchester, where he worked on modeling artificial intelligence. He died on June 7, 1954 under suspicious circumstances. More about his death in Section 5.4.

5.3 THREE PERSONAL TIDBITS: RUNNING, CHESS, HOMOSEXUALITY

5.3.1 Running

Alan Turing enjoyed running. There were times when he did it almost every day. Occasionally he participated in competitive races. He was a medium-to-long-distance runner. This included marathon runs, where he once came in fifth place.[4]

Here is what he wrote[5] to his mother in October 1946:

> My running was quite successful in August. I won the 1 mile and 1/2 mile at the NPL sports, also the 3 miles club championship and a 3-mile handicap at Motspur Park. [] The track season is over now, but of course the cross country season will be beginning almost at once. I think that will suit me rather better, though the dark evenings will mean that my weekday runs will be in the dark.

During this period he was running two to three hours every day.

This passion lasted throughout most of his adult life. For example, we are told that in 1951 to 1952 he may have run 1,000 miles cross-country together with a younger friend.

---

[4] [Hod], p. xxv; the number of participants is not given; it must have been at least a few dozen.

[5] Ibid, p. 434.

5.3.2 Chess

Henri Poincaré (more about him in the next chapter) wrote:[6] 'Every good mathematician should also be a good chess player and vice versa.' Turing was not an exception. It is not an overstatement to claim that Turing was a chess addict. When the 18-year-old Peter Hilton, who was to become a notable mathematician himself, joined the code-breaking team in 1942, he was greeted by Turing thus:[7] 'My name is Alan Turing. Are you interested in chess?' Chess sets were unavailable during wartime, so Turing made one out of clay.

Chess was not merely a game for Turing; it was a testing ground for his ideas on machine learning and artificial intelligence. He created perhaps the first ever chess-playing algorithm (that proved to be rather unsuccessful in practical games), and he daydreamed of a machine that would teach itself how to play chess. His dreams came true a few years ago: a self-taught AI machine handily beat the best chess-playing program, which in turn is superior to the best chess-playing humans.

A mathematician and the British chess champion of the time, Hugh Alexander, was Turing's friend, and he played a significant role in Turing's life. Alexander was drafted into the code-breaking team in 1941 and soon emerged as a coleader of the team. They, of course, played many chess games together. After the war Alexander continued playing chess. Although he never achieved the grandmaster title, he was a rather strong player.[8]

Alexander helped during the most difficult times in Turing's life. Which brings us to the next subsection.

5.3.3 Homosexuality

Turing was a homosexual. He was rather discreet about it, but he did not hide it either, even though he was aware that his society at large stigmatized homosexuality, considering it a perverse affliction. Turing found male bodies to be more attractive than female, public opinion be damned. From his friends, though, he demanded understanding and acceptance; here is what he told a good friend: 'We know each other quite well now ...I might as well tell you that I am a homosexual.'

---

[6] [Poi], p. 642.

[7] [Hil].

[8] In 1951, he managed to beat the grandmaster Svetozar Gligorić in a tournament game. Since I have played chess with many people who have played Gligorić, this establishes a short chain of three people connecting me chess-wise with Turing, and my *Turing-chess number* of 3.

In Turing's times homosexuality was illegal in Great Britain. The crime was that of 'Gross Indecency contrary to Section 11 of the Criminal Law Amendment Act 1885' and it was punishable by up to two years in prison. Turing considered the criminalization of homosexuality to be an anachronism, and he expected it to be repealed. This did not happen soon enough for him.

In 1951, Turing reported a minor theft in his house to the police. His frankness worked against him that time, as the police soon discovered that the perpetrator was his homosexual lover. Turing was summoned to stand trial. Eventually, with some help from his friends,[9] he escaped a prison sentence. As it turned out, what he ended up with may not have been much better: he was forced to undergo an experimental female hormone 'therapy' for a year. One side-effect of it was the enlargement of his breasts. However, he was not 'cured'.

In 1952, at about the time when the compulsory 'therapy' was over, he heard about men-only dances organized by the Scandinavian gay community in Norway, where homosexuality was not illegal. He took a trip to Norway and found no gay dances. On the other hand, he met and established contacts with 'five or six' gay men. His sexual life was back on track…

5.4 THE MYSTERY OF TURING'S DEATH: SUICIDE?

…So was his professional life: he held a stable position at the University of Manchester. According to people who worked with him, the intensity of his work had not diminished, and he was enjoying the research that he was doing. His social life was also normal: he often met with friends and acquaintances. He continued with his regular running or biking exercises. The difficult times of his trial and the forced hormonal injections were behind him. All in all, the last year of Turing's life was in general uneventful, as he settled into the quiet life of an established scientist.

So, his alleged suicide came about as a gross incongruency. On the Monday morning of June 7, 1954, he was found dead, lying neatly on his bed. By his side there was a bitten apple that had 'obviously' been laced with potassium cyanide. The apple was never analyzed, but that did not prevent the police from promptly declaring Turing's death as being caused by suicide.

---

[9] Hugh Alexander was one of the 'character witnesses' for Turing during his trial.

Nothing indicated that Turing had suicidal tendencies. A day earlier he was seen chatty and happy. He had reserved the university computer for the next day and he had theater tickets in his drawer. So, some, including Turing's mother, claimed that his death was an accident (negligence or Russian roulette).

Others claimed that he had been assassinated. This claim may not as preposterous as it may sound: documents show that, from the perspective of the Secret Service of the time, Turing's knowledge of military codes, his uncompromising frankness and honesty, and especially his homosexuality, combined with his frequent trips abroad, may have made him a liability (or 'security risk').

5.5 A MYSTIC OR NOT?

Here is what Turing wrote in 1932 in a letter to the mother of his deceased friend Cristopher:[10]

> Personally I think that spirit is really eternally connected with matter but certainly not always by the same kind of body. I did believe it possible for a spirit at death to go to a universe entirely separate from our own, but I now consider that matter and spirit are so connected that this would be a contradiction in terms. It is possible however but unlikely that such universes may exist.
>
> Then as regards the actual connection between spirit and body I consider that the body by reason of being a living body can 'attract' and hold on to a 'spirit', whilst the body is alive and awake the two are firmly connected. When the body is asleep I cannot guess what happens but when the body dies the 'mechanism' of the body, holding the spirit is gone and the spirit finds a new body sooner or later perhaps immediately.

Further, Turing 'believed that [the deceased friend] Christopher was still helping him.'[11] So, at 20, Turing comes about as a rational mystic.

In 1986, Arthur Hodges published a comprehensive, 776-page biography of Alan Turing (*Alan Turing: The Enigma*). It is our main source in this section. According to Hodges, Turing turned out to be an atheist and

---

[10] [Hod], p. 82–83.
[11] [Hod], p. 84.

a materialist all his mature life, but no quotes to back that claim up are provided. Hodges, as we will see, was strongly biased in that assessment.

In any case, in 1950, after remaining beyond the mystic horizon for a couple of decades or so, Turing emerged again as a rational mystic. That year he published an article titled *Computing machinery and intelligence*,[12] and in a subsection titled *The Argument for Extrasensory Perception* he wrote:

> How we should like to discredit [telepathy, clairvoyance, pre-cognition and psychokinesis]! Unfortunately, the statistical evidence, at least for telepathy, is overwhelming. It is very difficult to rearrange one's ideas so as to fit these new facts in. Once one has accepted them it does not seem a very big step to believe in ghosts and bogies. The idea that our bodies move simply according to the known laws of physics, together with some others not yet discovered but somewhat similar, would be the first to go.

Hodges's comment following this quote (page 524 in the *Enigma*) is telling:

> Readers might well have wondered whether he really believed the evidence to be 'overwhelming', or whether this was a rather arch joke.

We will come back to the 'joke' in the next section. Meanwhile, we can safely postulate that Turing was at least a rational mystic during the last years of his life. This conclusion is also backed up by the following intriguing anecdote that happened in 1954, less than a month before Turing died.

One Sunday in mid-May, 1954, Turing and his friends visited Blackpool. As they were taking a stroll by the seaside amusements, they came across a fortuneteller shop. Turing went in, and when he came out half an hour later, he was 'as white as a sheet'. He did not speak at all until they came back to Manchester.

The point here is not in his visit to the fortuneteller – that could be explained as being driven by his enormous intellectual curiosity. What is important is the fact that whatever the fortuneteller said – and one wonders if she had foretold of his impending demise – affected Turing profoundly. Atheistic materialists do not believe fortunetellers.

---

[12] [Tur]*.

5.6 THE 'JOKE'

In this section we briefly cover telepathy. It is usually defined as paranormal information transfer between the minds of beings, humans in particular.

Turing was certainly not the first modern scientist who expressed his displeasure with the shallow and facile denial and disregard of the telepathic phenomena by mainstream science. Konstantin Tsiolkovsky (1857–1935), widely considered to have been one of the founders of rocket science, wrote:[13]

> One cannot doubt the phenomena of telepathy. Not only is there a large accumulation of documents concerning these facts, but there does not exist a family whose members would refuse to testify to telepathic facts experienced by themselves. The attempt to explain these problems scientifically deserves our respect.

How similar to what Turing stated a few decades later!

By the way, Tsiolkovsky believed that 'humanity would reach its highest fulfillment through knowledge of space travel and parapsychology.'[14] He was way ahead of his time in many ways!

Confronting extraordinary deceptiveness requires stronger words than those written by Turing and Tsiolkovsky. The physicist Oliver Lodge comes closer to my taste when he wrote in his seminal work *Survival of Man:*[15] 'I venture further to say that persons who deny the bare fact [of the existence of the phenomenon of telepathy] are simply ignorant.' Arguably the most important psychologist of modern times, Carl Jung, came to a similar conclusion:[16] 'The existence of telepathy in time and space is still denied only by positive ignoramuses.' Ignoramuses are not the only ones who deny these phenomena. There are also those who know that the simplest way to manipulate humanity is through deceptive materialism and outright lies. This class of people is more ominous and disturbing.

So then, let's take a look at the part of the 'joke' concerning telepathy. Since this is not a book about paranormal phenomena per se, we must be brief: we will consider only two examples.

---

[13] [Vas], p. xiii.
[14] [But], p. 101.
[15] [Lod], p. 115.
[16] Quoted in [Fra], p. 149.

The biographer Hodges speculates that Turing must have been unduly influenced by 'J. B. Rhine's claims to have experimental proof of extra-sensory perception,' thus implying that Rhine's work was fraudulent. So, first we take a look at J. B. Rhine's research.[17]

Between the 1930s and 1950s, J. B. Rhine, mostly working in tandem with his wife Louise Rhine, conducted or supervised hundreds of thousands of experiments on telepathy, clairvoyance, and psychokinesis. A typical experiment on telepathy was conducted with so-called Zener cards (see the figure below). The *sender* would shuffle a deck of 25 Zener cards (five of each sort), pull out a random card, see it, then put it aside noting the time. A distant *receiver* would then try to guess that card. The chance probability that the guess would be correct is $\frac{1}{5}$, so that in a long sequence of runs of 25 guesses one would expect the average number of correct guesses to be around five.

Zener or ESP (Extra Sensory Perception) cards

An early subject of Rhine's experiments was Hubert Pearce, a divinity student. Pierce guessed all 25 cards in a single run; the odds of that happening by chance is $1:5^{25}$. How small is this number? Suppose one performs one such experiment of guessing 25 Zener cards by doing it one card per minute. Then a small computation shows that this event randomly occurs on average once in 14,175 billion years. The age of the 'universe', according to some, is 30 billion years!!

In 1936, during an invited talk about his results, a psychologist named Bernard Reiss effectively accused Rhine of being a liar. Rhine suggested to Reiss that he try a similar experiment. Reiss happened to be an open-minded skeptic, and in the fall of 1936, he started a telepathy experiment with Mrs. S, a woman of 26 who supported herself by giving music lessons. The outcome was nothing short of spectacular.

In each of the experiments, Reiss slowly flipped the cards of a shuffled deck of ESP cards, one card per minute. At precisely determined times when Reiss flipped the cards, Mrs. S, whose watch was synchronized with Reiss', made a guess of the card and noted it. At the time of the experiments,

---

[17] See, for example, [Rhi], [Rhi3], [Rao].

both of them were in their respective homes, a quarter of a mile away from each other. The expected number of hits per one run of 25 cards is around five. Indeed, five was precisely Mrs. S's score during her first run. But then it took off!

Between December 1936 and April 1937, Reiss and Mrs. S repeated the experiment 74 times until poor Mrs. S suffered a nervous breakdown. Here is the sequence of the numbers of hits in all of the 74 runs, ordered chronologically:

5, 7, 10, 12, 15, 8, 16, 13, 18, 21, 11, 15, 19, 24, 21, 21, 22, 24, 25, 24, 21, 20, 19, 20, 18, 14, 15, 15, 16, 12, 19, 21, 22, 24, 20, 18, 22, 21, 19, 19, 16, 21, 22, 17, 16, 18, 19, 21, 20, 19, 16, 21, 23, 19, 17, 22, 21, 18, 15, 14, 18, 18, 19, 18, 17, 18, 19, 20, 20, 20, 19, 20, 21, 21.

All in all, there were 1,850 trials, and the number of correct guesses (the sum of the numbers in the above sequence) was 1,349. Obviously, Mrs. S scored much, much better than the expected average. The *p*-value,[18] or the probability that one scores *randomly* 1,349 *or more* hits in 1,850 guesses of ESP cards, is 1.45×10^{-524}. This is about the same chance as winning the Lotto 6/49 jackpot (guessing 6 out of 49 numbers) 73 *consecutive* times by playing a single combination of 6 numbers each of the 73 attempts (worth at least $365 million)!

The readers should be aware that the very few results that we provide in this book are only in order to indicate the scientific context of the phenomena we have encountered or we will encounter. There are thousands of published articles on the so-called paranormal phenomena, based on literally millions of experiments. However, this is not a place for a more comprehensive review.

Scientific experiments statistically establishing clairvoyance will be briefly covered in the next section. Precognition, the third paranormal phenomenon mentioned in Turing's quote, will be briefly covered in Chapter 13. At this stage of our narrative I can only assure the reader that precognition is also genuine; its genuineness has been statistically established by numerous repeatable scientific experiments.

---

[18] Suppose an event happens with probability *p*; suppose we test it *n* times, and suppose we get *m* positive results. Then the *p*-value associated with this event is $\sum_{i=m}^{n} \frac{n!}{(i!)(n-i)!} p^i (1-p)^{n-i}$.

CHAPTER 6

A Gallery of Mystics and Mathematicians

First Digest

If we do not expect the unexpected, we will never find it.

HERACLITUS

In this chapter, we open the floodgates.

6.1 CLAIRVOYANCE

First, we elaborate further on Turing's 'joke'.

Clairvoyance is the paranormal information transfer between inanimate objects and the minds of humans. There is no sender in the case of clairvoyance, only a receiver. While telepathy can conceivably be explained away by postulating the existence of some hitherto unknown powerful, information-loaded brain-produced beams of energy, clairvoyance cannot. Hence accepting the genuineness of the phenomena of clairvoyance necessitates fundamental changes in the materialistic ontological paradigms.

Amazingly, it has been observed by many researchers that the information transmission rate in clairvoyance experiments is about the same as in telepathy studies: the sender is not really needed! One strange consequence is that explaining the ostensibly telepathic experiments as being

DOI: 10.1201/9781003282198-6

exactly that, namely cases of telepathy, might be off the mark. The deep reality seems much more interesting!

We start with the following delightful episode from the files of the Duke Parapsychology Lab researcher Margaret Pegram. In 1936, she experimented with children in a children's home. A nine-year-old girl named Lillian scored better than the others and was chosen for further testing. The test consisted of laying face-open a deck of 25 ESP cards, and Lillian would then try to match the cards in another (well-shuffled) deck (face down) with the open cards. Now quoting:[1]

> But then, [after scoring 3 and 8 in the first two attempts – the expected median score being, as we know, 5] she turned first slightly away from the table with a serious, 'Don't say anything, I am going to try something.' She closed her eyes for a few seconds, then, moving her lips as if speaking to herself, she slowly dealt out the cards. After she finished and before the cards were checked, she remarked, 'I was wishing all the time that I would get twenty-five.' She did. Every card was in the correct pile opposite its own key card.

Lillian wished for a miracle, and a miracle happened!

Hubert Pearce was mentioned in Section 5.6; he was the receiver in one of Rhine's telepathy experiments. He told Rhine that he believed he had inherited his mother's clairvoyant powers. So they tested his clairvoyance, and indeed, as it turned out, he didn't need a sender to correctly guess the cards.[2] In the early 1930s, during a period of 18 months and under the supervision of Rhine's assistant J. G. Pratt, Hubert participated in nearly 700 runs through the standard deck of 25 ESP cards; his average success rate was 32% (the mean chance being 20%). The p-value of this result is only 10^{-305}, which is about the same as winning the 6/49 jackpot 43 consecutive times, by playing a single combination each time!

We fast forward about 30 years for our last batch of clairvoyancy experiments. In the mid 1960s, Czech biochemist and researcher Milan Ryzl experimented with a mild-mannered clerk named Pavel Štepanek.[3] Ryzl would hypnotize Štepanek who would then try to guess the color of a card

---

[1] [Rhi], p. 46–71.
[2] [Rao], Chapter 2.
[3] [Ryz]*, [Dup], [Ost].

put in an opaque envelope. There were only two colors, so the probability of a correct guess was one out of two (or 0.5). In 5,000 trials Štepanek got 3,611 correct answers. The odds this happened by chance, or the p-value of the event, is less than 5.3×10^{-648}. That is about the same as the chance of winning 90 consecutive jackpots of Lotto 6/49 with single combinations ($450 million; tax-free)!! In the subsequent trials, Štepanek self-hypnotized himself, and his success rate somewhat decreased: he scored 11,978 hits in 19,350 binary calls (Ryzl, 1966). The probability of this happening randomly is 6.39856×10^{-243}, about the same as the chance of winning 34 consecutive jackpots of Lotto 6/49.

The implications of these and other results are staggering. As Turing put it, 'ghosts and bogies' are not far away. Consequently, such results have been consistently swept under the carpet and thoroughly ignored by the materialists. However, as we will see in the next section, the Ryzl–Štepanek case is an exception.

6.2 MARTIN GARDNER: A DEBUNKER

One fine day many years ago, when I was young and free, I decided on the spur of the moment to hop into a train and travel 450 km to Belgrade. It was an easy decision: the fare was cheap, and at that time I had about a dozen uncles, aunts, and cousins living in Belgrade, so my stay there was free. And I had a clear reason for going there: I wanted to visit the newly-opened 'bookstore with books in foreign languages'. There I bought my very first mathematical book in English: *Mathematics Magic and Mystery*[4] by Martin Gardner. I enjoyed it thoroughly!

Martin Gardner (1914–2010) was not a professional mathematician; however, he had an undeniable mathematical talent, an appreciation of the beauty of mathematics, and a flair for writing. All of these earned him a popular long-standing column in *Scientific American*. *Scientific American* being a magazine of the scientific establishment, it certainly helped that he was an ardent materialist and a devoted denier and 'debunker' of everything 'paranormal'.

Once, in 1974 or soon after, Gardner claimed in one of his *Scientific American* columns that he was 'not guessing' when he accused the prominent psi-researcher (= paranormal-researcher) Russell Targ of sifting

---

[4] I realize now, as I write this, how uncannily similar Gardner's title is to the title of the book I am writing! In fact, staying on our anti-chronological course, Gardner simply translated my title of some 65 years later into the language of materialism: mathematics -> mathematics, miracle -> magic, mysticism -> mystery!

out the negative or unsuccessful data in an ESP study carried out at the Stanford Research Institute (SRI) and partially funded by NASA. Years later he confessed to a reporter that he had made it up, 'because that's the way it must have been.' This reporter, according to Russell Targ,[5] was so shocked that he felt compelled to call up the SRI research team and share with them Gardner's admission.

Martin Gardner, ~2005

We digress for a paragraph: *Scientific American* has been a long-standing platform for the defenders of dogmatic materialism, particularly 'magicians'. For example, in 1924, they assembled a committee with a specific task of discrediting mediums (more about mediums later in this chapter). One of the members of the committee was the magician and debunking crusader Harry Houdini.[6] The committee expressly dispatched the first three mediums. However, the fourth one, the psychic Margery Crandon, produced unexplainable poltergeist-like phenomena. And so, as it has been asserted,[7] Houdini and his assistant planted a ruler in the specially designed cabinet in order to discredit the psychic. Said this assistant: 'There's one thing that you've got to remember about Mr. Houdini – for him the truth was bloody well what he wanted it to be.' Indeed, Margery Crandon was declared fake as well.

The a priori negative attitude regarding everything paranormal is typically the first and largest step in the debunkers' argument. The rest of it is usually not influenced by the experiment itself but by the debunker's (professional) background: for example, 'scientific skeptics' search for or

---

[5] [Targ], p. 157.
[6] Originally Erich Weiss. He gave himself the name after the 19th-century magician Jean Eugène Robert-Houdin; unlike Houdini, the original Robert-Houdin accepted the general genuineness of the mediumistic phenomena.
[7] [Wil], p. 232.

imagine loopholes in the protocol (once they claimed that 'there was a crack in the door through which the subject must have surreptitiously peeked'), magicians invoke various tricks.

When Martin Gardner learned of the Ryzl-Štepanek experiment, he knew immediately it was a fake because 'that's the way it must have been.' Being an amateur magician himself, he also knew what happened: Štepanek fooled Ryzl through some sleight-of-hand tricks. So certain was he in his verdict that not only did he publicly debunk the experiment but he also offered Štepanek money if he would admit the deception.

Gardner was wrong. Štepanek was singularly incapable of doing sleight-of-hand tricks: he had suffered from a condition called webbed fingers, and his dexterity was minimal during his adult life. Moreover, he did not handle the envelopes at all! The cards were in opaque envelopes that were in turn in book flaps, and Štepanek handled them as instructed, by holding only the edges of the flaps for a couple of seconds. This was also checked by a group of Dutch researchers. One more thing:[8] 'The thorough Dutch [scientists] also asked a world champion magician, Fred Kaps, if he could duplicate Pavel's card calling, under identical conditions, with his magic. Kaps, at length, told them his bag of tricks couldn't match Stepanek's ESP performance.'

The above critique of the Gardner-debunker does not diminish at all the fact that he was a great writer and popularizer of mathematics. I still appreciate and enjoy his work.[9] However, he plays a negative role in our narrative. The only such role!

6.3 SAMUEL SOAL AND ELEANOR (NORA) SIDGWICK

Both were presidents of the (London) Society of Psychical Research (SPR): E. M. Sidgwick 1908–1909 and 1932–1933, and S. G. Soal 1950–1952. There are at least two more instances of mathematicians who were presidents of the SPR: George N. M. Tyrrell (1879–1952) and Sir Oliver Lodge (1851–1940); both were physicists as well.

6.3.1 Samuel George Soal

Samuel George Soal (1889–1975) was a mathematician and a lecturer at Queen Mary College, University of London, his alma mater. He was also

---

[8] [Ost], p. 341.

[9] Second coincidence regarding Gardner: I just realized by searching the internet that I have been published together with Martin Gardner in the same issue of *Mathematical Intelligencer* (2001, Vol. 23, Number 1); twice, if we count the review of his autobiography (2014, Vol. 36, Number 1).

a parapsychology researcher, mostly examining the 'telepathy' phenomenon through experiments of the sort we have encountered above.

In the period from 1936 to 1941, he tried to replicate Rhine's results: in a run of experiments involving 160 participants who were tested in 120,000 Zener-card guessing trials, the cumulative results were at about chance odds. The subsequent results were strongly positive.

Samuel Soal (middle)

Soal published two books. In the second, titled *The mind readers: Some recent experiments in telepathy* (1959, in collaboration with H. T. Bowden),[10] he gives the details of telepathy experiments with two Welsh boys. The boys, Glenn and Ieuan, both about 15 years old, were cousins from a remote North Wales village. Soal purposely chose a corner of the Celtic world where the rich 'paranormal' lore of bygone ages still lingered, including the conviction that telepathy, under a different name, is a normal phenomenon.

Soal used a children's variant of Zener cards: instead of the abstract symbols, there were drawings of a lion, a penguin, a giraffe, an elephant, and a zebra. Ieuan was the sender, Glenn the receiver. The cousins performed well above the chance odds: in 15,348 trials there were 5,461 correct guesses, where the mean chance expectation was 3069.6. The odds that the event happened by chance (the *p*-value) is about 10^{-438}. Using a comparison other than Lotto 6/49, this is about the same as getting face-up for *every* coin after throwing a pile of 1,455 coins in the air. Still, this is far from being a singular result, even in the context of the very few examples that we have mentioned earlier.

---

[10] [Soa].

There is a very interesting episode related to Soal's experiments that illustrates the unhinged ferocity of the defenders of orthodox materialism. The families of the two boys were poor, and the financial incentives given to the boys for their participation in the experiment meant a lot to both families. The long-running experiments became duties for the boys and were burdensome to them, especially during the days when they much preferred playing outside to sitting still for hours on end. So, it is not surprising that they were once caught cheating: they attempted surreptitious auditory clues through coughing, making noise with the chair, or similar. Soal told the parents that the behavior of the boys was 'very unsatisfactory,' and, in the spirit of their times, the fathers gave them 'a good dressing down'. After this incident, Soal straightened the controls, and the results of the experiments during the month when the cheating occurred were removed from the final tally.

For the skeptical reviewers that was not good enough. According to them, the whole years-long experiment must have been fraudulent, the main argument presumably being the ubiquitous, 'because that's the way it must have been.' One of the skeptics even imagined that the two boys had used an ultrasonic whistle! In another case, a 'scientific skeptic' speculated that Soal was engaged in 'unconscious whispering'! Could the accusations have been more frivolous? Perhaps. However, it is difficult to surpass the disingenuousness of a prominent online encyclopedia whose main *other* purpose is to trace for us the thin lines of *proper* or sanctioned thinking, in just about everything. In the case of Soal's entry, they set the tone right off the bat: 'He was charged with fraudulent production of data in his work in parapsychology.' In other words, Soal's experiments, for what they were worth, are to be ignored, or else you're not *scientific*. Case closed.

6.3.2 Eleanor-Nora Mildred Sidgwick

Eleanor-Nora Mildred Sidgwick, (1845–1936) was born into one of the most prominent 19th-century British families. She and her siblings were given excellent in-home education: her brother Arthur Balfour became the Prime Minister of the United Kingdom (1902–1905), and her sister Alice Blanche was a pioneer in the study of genetics. Nora studied mathematics and eventually became a mathematics lecturer (and principal) at Newnham Women's College, Cambridge. She has been referred to as[11] 'one of the most powerful intellects of her generation'.

---

[11] [Spo]: article by John Hick starting on page 93.

Nora Sidgwick

She was certainly a very thorough psi-researcher. Consider, for example, the article she published in the *Proceedings of the SPR* called 'A Contribution to the Study of the Psychology of Mrs. Piper's Trance Phenomena', 1916. The article is a study of 657 pages, that some people[12] call 'one of the most important contributions that we have to parapsychological literature'.

She was also the principal investigator in a study on hallucinations and the main writer of the subsequent report: 'Report on the Census of Hallucinations' by H. Sidgwick, A. Johnson, F.W.H. Myers, F. Podmore and E.M. Sidgwick (1894, Proceedings of the SPR).[13] The 392-page article was based on the data collected from 17,000 individuals, who were asked the following question: 'During waking life, have you ever heard a discarnate voice, seen a vision of somebody who had died, or perceived any sensation that seemed to have no physical cause?' She got affirmative answers from 1,684 people. Of these, 32 were visions of people who had died, but the perceiver had not been aware of their death!

Eleonore Sidgwick goes down in our book as at least a rational mystic.

By the way, the first listed author of the 'Report on the Census of Hallucinations', the philosopher Henry Sidgwick, was Nora's husband, also a prominent psi-researcher and the founder of the Society of Psychical Research. For a substantial period of time, he was also the president of a close-knit social club, gathering some prominent Cambridge intellectuals;

---

[12] Ibid.
[13] Summary in [Mac], p 8

one of the members was Alfred North Whitehead, a subject of the following section.

6.4 ALFRED WHITEHEAD AND BERTRAND RUSSELL

We remind the reader: as per Turing, it is a short distance from clairvoyance to 'ghosts'. From the latter to post-mortem non-corporeal existence there is no distance at all. We have used shortcuts, and we have selected just a few examples from the abundant scientific evidence justifying the genuineness of clairvoyance. There is no choice: elaborating fully would take us too far from our subject. We will continue selecting some supporting results, going in the same direction, until we reach the point of no return. Along the way, I will try to minimize referencing the huge library of published anecdotal evidence, focusing mainly on controlled laboratory results.

6.4.1 Alfred North Whitehead

Mathematicians remember Whitehead and Bertrand Russell primarily for writing *Principia Mathematica* (1911–1914). The main purpose of the multivolume *Principia Mathematica* was to establish an axiomatic framework for formalizing mathematical propositions in the language of symbolic logic. The project was abandoned after Volume 3. However, the *Principia Mathematica* publication is a huge act in the history of mathematics that profoundly influenced the development of mathematics in general. It was Whitehead's (and Russell's) main mathematical contribution.

Alfred North Whitehead

A. N. Whitehead (1861–1947) was also a distinguished philosopher, especially during the second half of his life. Judging solely from the secondary bibliography, it seems that he was more influential as a philosopher than as a mathematician: up until 1977 there were 1,762 books or articles *about* A. N. Whitehead, and almost all of them are about Whitehead-the-philosopher.[14]

The profound shift in his interests, professional life, and beliefs (from atheist to theist) closely parallels Gödel's. According to Russell and Whitehead's children,[15] the change was brought about by a personal tragedy: his younger son was killed in action during the First World War. That may have been the main cause, but I strongly suspect that there was another, related trigger. This speculation is not based on the available sources – almost all of Whitehead's personal notes and letters were destroyed as per his will – but rather on psychological considerations and statistical evidence. One does not embark on creating elaborate philosophical schemes in order to deduce immortality as a response to bereavement, as Russell implied. It is necessary to either have a direct personal experience that could be interpreted as implying a kind of immortality, or to accept and empathize deeply with someone else's experience of the same sort.

In the former case, the experience may have been a vision (or a convincing hallucination). This is not a far-fetched claim: as we mentioned earlier, according to many sources, the majority of people experience some kind of nonpathological hallucination at least once during their lifetimes. This percentage is much larger in the case of bereaved spouses or parents.[16] So, in that sense, the hypothesized event in Whitehead's life is more ordinary than singular. Nevertheless, dogmatic materialism, disguised as scientific skepticism, attaches stigma to occurrences such as visions, which in turn pushes the events into the social underground. From the times of Pascal, some four centuries ago up to the present times, the experiencers, especially those who resented notoriety (Whitehead, Gödel, etc.), kept quiet and diligently hid their tracks.

In the alternative scenario, Whitehead's initial acceptance of immortality may have been influenced by other people. One man stands out: Oliver Lodge, whom we mentioned at the beginning of the previous subsection.

---

[14] [Woo].

[15] [Low], p. 188.

[16] [Star], p. 88.

Lodge (1851–1940) was a prominent (mathematical) physicist and inventor. His personal life parallels that of Whitehead's: Lodge's son Raymond was also killed in the First World War. That happened in 1915. Immediately after that Lodge witnessed paranormal phenomena, convincing him that his son, or his consciousness, still existed. He chronicled his experience in the book *Raymond or Life and Death*.[17] This was a widely read book at the time, and Whitehead must have read it too.

I must admit that at times I had difficulties extracting clear meaning from Whitehead's philosophical writings. I consider this to be primarily my fault: I never liked words, especially neologisms defined through notions that are obscure to me. Here is an example:[18] '["Everlasting" is] the property of combining creative advance with the retention of mutual immediacy'. The fact that I am not the only one[19] who found it difficult to understand Whitehead's philosophy does not diminish my shortcomings.

We give two examples of Whitehead's mystic-like claims that I have extracted from my reading of his work. Quoting from his preface to *Process and Reality*:[20]

> [Relatedness of actualities] is wholly concerned with the appropriation of the dead by the living – that is to say, with 'objective immortality' whereby what is divested of its own living immediacy becomes a real component in other living immediacies of becoming.

That is, he is saying that what we call death is merely a change in the state of living. This has often been claimed by true mystics, who in addition often assert that there are seven 'spheres of existence'.

He also wrote 'The world is in the soul,'[21] which is unequivocally mystical. Indeed, the Neoplatonic philosopher and true mystic Plotinus (204–270) stated some 1,800 years before Whitehead that 'Every being contains within itself the entire intelligible world.'

---

[17] [Lod2].

[18] [Low], p. 246.

[19] In 1927–1928, Whitehead gave a sequence of ten lectures in Edinburgh. The first one was attended by about 600 people, the second by about six, and the last one by two. Source: Lowe: *Alfred North Whitehead: The Man and His Work*, p. 250.

[20] [Whi].

[21] [Whi2], p. 224.

We will close our brief account on Whitehead: he was a kind of tentative (rational) mystic. Given what we know about him, as opposed to our speculations, he is possibly our weakest case.

6.4.2 Bertrand Arthur William Russell

Bertrand Russell (1872–1870) was not only a prominent mathematician; he was also an eminent philosopher and writer. In 1950, he was awarded the Nobel Prize in Literature. Unlike Whitehead, Russell had an exuberant personal life. He married four times, divorced three times, and he fell in love with Whitehead's wife to boot. In 1918 he was imprisoned for speaking out against the First World War; Arthur Balfour, Nora Sidgwick's brother mentioned earlier, intervened to change his sentence from Second Division (ordinary convicts) to First Division (convicts with special privileges). When he was 89, he was imprisoned again, this time for 'breach of peace': he participated in an anti-nuclear demonstration. He was 98 when he died.

Bertrand Russell

Was the man a mystic? Not by any stretch of imagination! He was a hardcore, atheistic materialist. He wrote in the essay *A Free Man's Worship*[22]

> that [man's] origin, his growth, his hopes and fears, his loves and his beliefs, are all but outcomes of collocation of atoms [], are things, if not quite beyond dispute, are yet so nearly certain that no philosophy which rejects them can hope to stand.

---

[22] Quoted in [Lod3], p 7

So then, why is he included? Because 'he' changed his mind. And that happened a year after he stopped being a man! That is, after he died!

We remind the reader again that the orthodox materialistic rules-axioms of the 'reality' do not apply if one accepts – as I do – the overwhelming statistical and anecdotal evidence confirming the genuineness of clairvoyance and other 'paranormal' phenomena. The paranormal becomes normal.

Russell's categorical proclamation notwithstanding, there is tons of evidence proving that (human) consciousness can exist incorporeally. According to the prominent psychiatrist Stanislav Grof, it is remarkable how academic circles have managed to suppress or ignore this evidence.[23] This incorporeal existence is indisputable;[24] the problem is in the human inability to perceive outside a minuscule interval of the light spectrum. Nevertheless, there are talented individuals, sometimes called mediums or channelers, who *can* perceive and communicate with discarnate or disembodied entities. This is also a result confirmed in controlled experiments.[25]

Channelers are special mediums: they let their bodies, specifically their vocal cords, be used by incorporeal entities.[26] And so it happened that the hypothetical Bertrand Russell (HBR) manifested a year after he died, through a school cook called Rosemary Brown.

Here is what HBR said (through Rosemary) during his first appearance:[27]

> You may not believe that it is I, Bertrand Arthur William Russell who am saying these things, and perhaps there is no conclusive proof that I can offer through this somewhat r1estricted medium. Those with an ear to hear may catch the echo of my voice in my phrases, the tenor of my tongue in my tautology; those who do not wish to hear will no doubt conjure up a whole table of tricks to disprove my retrospective rhetoric.

And of course, compared with BR's, HBR's metaphysical philosophy was substantially different:

---

[23] [Til], p. 3.

[24] Here is one of many examples: in an experiment performed by Gary E. Schwartz, the entrance of disembodied entities in a completely dark chamber was registered via 'Fast Fourier Transformation' image analysis. More about this experiment is in Section 7.4.

[25] More about this in Section 8.4.

[26] A good book about channeling is Jon Klimo's *Chanelling* ([Kli]).

[27] [BroR]; also in [Las], p. 151.

I am far less a cynic than I was []. Do I believe in God now? Many people will want to know my answer. Yes, I now believe without equivocation, and with a positive intellectual comprehension which was and is the sole acceptable proposition as far as I am concerned.

Further: '[Men] are gods in the making, but the making has to be of their own. The tragedy is that so few make it.'[28] Are these words of HPR, or of a cook who worked in a school kitchen?

We return for a moment to Chapter 3 where we mentioned the following quote, attributed to W. H. Myers: 'Freedom can only come through the deliberate process called "The Breaking of the Image."' As was the case with HBR, this message was also delivered by hypothetical (dead) Myers through a medium![29] However, there is a major difference: the hypothetical Myers' antecedent was the living Myers, who, as we have mentioned in the previous section, was a prominent parapsychological researcher during his lifetime. So, the hypothetical Myers went much further: he participated in a cross-correspondence experiment when he delivered two meaningless phrases to two distant mediums simultaneously so that when the phrases were put together, the result was a coherent and meaningful message.[30]

Bertrand Russell was not a mystic, but HBR was a mega mystic, if this attribute applies at all in such cases.

Ironically, there was a hypothetical Harry Houdini (HHH) too! First, we note that, despite professing that he had not witnessed a single genuine psychic manifestation, Houdini told the writer and psi-investigator Hereward Carrington that he had seen an apparition of his mother while performing in Berlin, at the time when his mother had been dying in New York.[31] Since he took the event at face value, there was at least a desire to believe in post-mortem existence. Houdini died in 1926. A few days earlier, he had told his wife that he would try to manifest and produce a certain message in the code known only to the two of them that they used in fake telepathy performances. HHH manifested through several mediums, phony or genuine. In 1927, the medium Arthur Ford produced the following jumbled message from HHH: 'Rosabelle, answer, tell, pray-answer, look, tell, answer-answer, tell.' The message was delivered to Houdini's

---

[28] The great mystic and poet William Blake said essentially the same thing.
[29] [Cum], p. 67.
[30] Ibid.
[31] [Car], p. 61.

wife who decoded it to 'Rosabelle, believe.' That was what she expected. The relevant part of the code was: *answer* = B, *tell* = E, *pray-answer* = L, *look* = I, *answer-answer* = V.

6.5 JOSEPH WHITEMAN AND OBE

An out-of-body experience (OBE) is the sensation of visual perception from points of view external to the body of the percipient. It usually starts with what in medicine is called autoscopy, perceiving oneself from the outside. We will, in general, not follow the medical-pathological interpretation of OBEs. Neither we will call them dreams or illusions. There are many reasons for making that choice. We will briefly describe just a few scientific experiments strongly supporting our choice.

Miss Z. had an unusual talent: she could induce OBEs while sleeping. One might say, 'well, if she was sleeping, she must have been dreaming'. As we will see, there are at least two reasons why this argument is false.[32]

In 1965, the psychologist and psi-researcher Charles Tart suggested to Miss. Z that she write the numbers from 0 to 9 on pieces of paper, put the papers in a box, choose one piece randomly, and place it on the bedside table without seeing the number before going to bed. Then she would try to see that number during an OBE. Miss. Z followed the instructions for nine consecutive nights and subsequently reported to Tart that she *saw* the number correctly each of the nine times (chance odds 1 out of 362,880).

Tart suggested a similar experiment under controlled conditions, and Miss. Z agreed. She slept alone in a laboratory, where she was monitored through an observation window for four consecutive nights. Tart instructed Miss. Z to try to identify the five-digit random number he wrote on a piece of paper that he secretly put on top of a high shelf. During the last night Miss. Z, who was angry at herself for not being able to perceive the number (despite multiple OBEs during the first three nights), managed to experience a ten-minute long, early-morning OBE, when she succeeded in identifying the target number (25,132). The chance that she did it by sheer luck is 1 in 90,000. She couldn't stand up without being noticed since that would have disrupted the functioning of a brain-wave-recording machine in the other room to which she was attached. What is also very interesting is that the electroencephalogram (EEG) showed altered brain waves and no REM (rapid eye movement) during the period Miss. Z had had the OBE. Since REM is closely linked to dreaming, this almost excludes explaining away the OBE as being a dream. At the end

---

[32] [Tart]*; see also Tart's article in the collection [White] (*Psychic Exploration*), p. 349 to 374.

of this digression, we note without elaborating[33] that experiments with the talented psychic Ingo Swann, performed by many researchers, show that telepathy is not a feasible explanation of the phenomena of acquiring information through OBE.

This is by far not the only controlled experiment demonstrating the genuineness of extrasensory perception during OBEs. An OBE in itself is a strong indication that consciousness can exist incorporeally. Indeed, there exists ample anecdotal evidence implying that the default domain of consciousness is outside the realms determined by time, space, and matter. In any case, we will concur with the view of a number of the mathematicians featured in our narrative and accept the hypothesis that (human) *consciousness can exist independently of (what we call) matter.* This is a momentous postulate with profound consequences. So, we provide a bit more scientific support.

There are many OBE reports during attempts to resuscitate victims of heart strokes. The patients would find themselves out of their bodies, they would then calmly observe the procedure from a vantage point close to the ceiling and afterwards describe details plus some specific idiosyncratic moments. The standard objection of the debunkers would then be centered on the hypothesis that people's descriptions are based on their earlier knowledge. This was in turn debunked by the cardiologist Michael B. Sabom. Sabom investigated 32 cases of people giving autoscopic (self-observing) descriptions during near-death experiences (NDE) related to surgery or other medical procedure. Of these, none made any flagrant error in describing the CPR (cardiopulmonary resuscitation) that they had observed during the ensuing OBE. He had a control group of 25 people whom he asked to describe what they thought had happened during their own CPR. Of these people (none of which had NDE), 20 made cardinal errors, three gave limited correct descriptions (at least one of them had seen the procedure done on his father), and two did not know. Of the subject group of 32 cases, there were six of them who gave very detailed and correct descriptions of the procedure. These six are described in detail in Sabom's book *Recollection of Death: A Medical Investigation.*

The overall picture suggested by these experiments and results is stunning: the possibility of the incorporeal existence of consciousness becomes not a matter of belief; rather, the main issue is the choice between acceptance and denial of the overwhelming statistical and anecdotal evidence establishing the genuineness of the phenomenon.

---

[33] See, for example, [Cur], p. 130.

We use our short OBE sojourn to set up the stage for our next entry: Joseph Hilary Michael Whiteman (1906–2007).

There is nothing tentative about the mysticism of Joseph Whiteman: he journeyed the mystic roads extensively. He was a prolific 'out-of-body' traveler. He had more than 3,000 episodes of OBE, end it seems that he meticulously noted all of them in his diary.

Joseph Whiteman

Whiteman was born in England where he received his education, all the way to a degree in mathematics. In 1937 he emigrated to South Africa. In 1944 he was awarded a PhD (on the continuum hypothesis[34]) by the University of Cape Town. Between 1944 and 1946, he worked as a lecturer in music (!), and from 1946 until he retired in 1972, he was a lecturer, then associate professor of applied mathematics at the University of Cape Town. He died in February 2007,[35] aged 101.

He started with a bang: his first OBE took him all the way to an ecstatic union with the Supreme Consciousness. It happened in 1932 when he was 26. Here is an insert:[36]

> At once, without any further change, my eyes were opened. Above and in front, yet in me, of me, and around, was the Glory of the Archetypal Light. Nothing can be more truly Light, since that Light makes all other lights to be light; nor it is a flat material

---

34 A reduced version of the continuum hypothesis reads as follows: there is no set of cardinality (= the 'number' of elements) strictly between the cardinality of the set of integers and the cardinality of the set of all real numbers.

35 Coincidentally, I started reading his book *The Mystical Life* only a couple of weeks before he died.

36 [Whm].

light, but a creative light of Life itself, streaming forth in Love and Understanding, and forming all other lives out of its substance.

He achieved a state of mind sometimes called cosmic consciousness; it is the last stop in mystical journeys.

At that time Whiteman was not familiar with out-of-body phenomena. Nevertheless, they kept coming: by 1955 he had accumulated and noted more than 2,700 OBEs, and entries in his diary. Whiteman makes a clear distinction between dreams and OBEs: the latter start with a separation of consciousness from the physical body, and they have other features that dreams do not. For example, many times he perceived his physical body from outside, and a few times he had a double consciousness: the sensation of having two points of view simultaneously, one from the body and the other from outside. Here is one such case (experience #422, year 1934):

At the conclusion of a short separation of fantasy type, not otherwise noted, I was resting on open ground in a rather bright light when the need for return was recognized, the state being at that time relatively free of fantasy. I appeared to be brought back so that the conscious form came into coincidence with the physical form, but now so that physical space was finally confirmed. The bright light and the open ground in the one space could be perceived, at the same time as the body appeared in bed in the other space. Moreover, by the exercise of a free choice and decision, the former space could be more and more confirmed, so that the latter faded away, and once again I appeared to lie on the open ground in a bright light. And again, by relaxing the intention, the space as of the bedroom could be gradually restored, while the other space faded away. In fact, the two spaces were held for a few seconds on perfect balance, both being as if present simultaneously, and twice the space of the open ground and bright light was voluntarily restored before a final confirmation of physical space was made.

Said the wise shaman Don Juan:[37] 'There are worlds upon worlds, right here in front of us.'

Whiteman published five books on the subject of mysticism; the above cases are taken from his first book, *The Mystical Life*, 1961.

---

[37] [Cas].

6.6 HENRI POINCARÉ AND GEORG CANTOR

6.6.1 Jules Henri Poincaré

In 2006, I attended the International Congress of Mathematicians in Madrid. As is always the case with large meetings, the presence of almost everybody among the 4,500 participants was not noticed by almost everybody. However, everybody eventually noticed the *absence* of one person: Gregory Perelman. He had not come, refusing to accept the one million dollar Millennium Prize for solving in affirmative the *Poincaré conjecture.*

In simple terms (sacrificing rigor), the Poincaré conjecture states that the only bounded,[38] connected 3-manifold (= a space that locally *feels* like the *ordinary* three-dimensional space) that doesn't have any holes is the three-dimensional sphere. The three-dimensional sphere can be obtained from a ball (= the usual two-dimensional sphere, together with the points inside the sphere) by identifying – in your mind – every point in the northern bounding hemisphere with the point in the southern bounding hemisphere which is straight below it. In order to indicate what we mean by 'holes', we point out that the usual sphere doesn't have any holes, while the surface of a donut does, and that the former is the only hole-less, bounded, connected 2-manifold.

The Poincaré conjecture was named after Henri Poincaré (1854–1912), one of the greatest mathematicians ever.

Henri Poincaré

---

[38] compact.

We only have sporadic indications of Poincaré's rational mysticism. However, it seems safe to postulate that Poincaré agreed with some of the mystic tenants. Here is an example: *A reality completely independent of the spirit that conceives it, sees it, or feels it, is an impossibility.*[39] If the last desert tree fell and if there is not a single creature around to hear it hit the ground, was there sound at all? No, according to Poincaré. That is very mystic-like! So is the following claim that what we call 'reality' is in fact our construct:[40] 'In order to classify some phenomena in a simple way, it is most convenient to assume that the earth rotates.'

We also have the following Neoplatonic statement:[41] 'All that is not thought is pure nothingness....'. And the following: 'The subliminal self is in no way inferior to the conscious self. It knows how to choose and to divine.'

Finally, here is a curious anecdote that may be apocryphal:[42]

> one night he [Poincaré] retired to rest after thinking deeply on the problems for a long time, and on getting up the next morning he discovered to his intense surprise on his table several sheets of paper on which he had worked out a complete solution of the problem.

In 1884 Poincaré and the mathematician Georg Cantor met in Paris. They talked mathematics and established a good rapport: Cantor wrote in a letter to a friend that he liked Poincaré very much.

6.6.2 Georg Ferdinand Ludwig Philipp Cantor

Note: the next paragraph contains 'skippable' mathematics.

Given a set X, we denote by $P(X)$ the set of all subsets of X. For example, if $X = \{1, 2\}$, then $P(X) = \{\varnothing, \{1\}, \{2\}, \{1, 2\}\}$. If X is finite, then it is clear that the number of elements in $P(X)$ is larger than the number of elements in X. To the surprise of many, it was shown by the end of the 19th century that the analog statement is true for infinite sets too. Let's see why. (No prerequisite is necessary in order to understand what follows; only concentration.) Suppose X and $P(X)$ are of the same size. Then we can

---

[39] *The Foundations of Science: Science and Hypothesis, The Value of Science, Science and Method,* translations of Poincaré's works 1902, 1905, and 1908, University Press of America, 1982. (See https://plato.stanford.edu/entries/poincare/.)

[40] Quoted in [Dal], p. 94.

[41] *The Value of Science*; Chapter 11: *Science and Reality*, 1907. Book by Henri Poincaré, translated by George Bruce Halsted.

[42] [Joh], p. 46; original source not given

put all of the elements of X into one-to-one correspondence with all of the elements of $P(X)$. We denote by S_a the set in $P(X)$ that is associated with the element a in X under this correspondence. Construct a subset Z of X as follows: if an element a of X does NOT belong to S_a then we put it in Z; otherwise we don't. This set Z belongs to $P(X)$; so, there must be an element z in X, such that $Z = S_z$. Is z in Z? It follows immediately from the way we have constructed Z, that if z is in Z then z is NOT in Z, and that if z is NOT in Z then z is in Z. Our assumption that X and $P(X)$ are of the same size led us to a contradiction. So, the size of $P(X)$ is always larger than X. When X is infinite, this tells us that there is a hierarchy of infinite sets with regard to their sizes: there are different 'numbers' of elements in infinite sets, giving rise to *transfinite numbers.*

The discovery of transfinite numbers is due to Georg Cantor (1845–1918). It was a controversial concept at the time.

Georg Cantor

Cantor was born in St. Petersburg, Russia, to an affluent and cultured family. His father was a businessman and his mother a musician. When he was 11 years old, his family moved to Germany. After finishing Gymnasium (high school) in 1863, he moved to Berlin, where he received his university education. He got his PhD in 1565, and a couple of years later he joined the University of Halle as a privatdozent. He was only 34 years old when he became a full professor.

Despite this success, throughout his adult life, Cantor was haunted by lingering doubts about whether he had made the right choice by choosing mathematics over music. He had inherited a natural talent for music from his parents and was a talented violinist. Cantor hoped to relive the life choice that he did not make through his musically talented youngest son. That came to a tragic end: his son died unexpectedly in 1899, four days before turning 13.

During the last 25 years of his unhappy life, between 1884 and his death in 1918, Cantor suffered from mental breakdowns (manic depression and paranoia), and he was confined to hospitals and sanatoriums (then commonly called lunatic asylums) for longer and longer periods. Some aspects of his professional life may have triggered the bouts. He was sensitive and very upset by unfriendly mathematical critique. Cantor was also frustrated by his inability to prove the continuum hypothesis (mentioned in footnote 34, Section 6.5).

Regarding the continuum hypothesis, we mention that it was eventually proven that it could neither be proven nor disproven (within the standard set theory). The former result was achieved by Paul Cohen (1963), and the latter was proven earlier, by none other than Kurt Gödel (1940). So, Cantor tried for years to prove something that was not provable. This is a *curse* in mathematics, and many a mathematician has wasted their lifetime trying to do something that was eventually shown to have been undoable. A prime example is the hope that the fifth Euclidean postulate – that there is a unique line parallel to a given line and passing through a given point – is a consequence of the other four axioms of Euclidean geometry.

Cantor was religious: he was Protestant. He was a mystic too! Like Gödel and Whitehead, Cantor supplanted his focus on mathematics with metaphysics – in his case mostly with Christian philosophy – during the last part of his life.

He wrote in 1913:[43] 'Everything we perceive with the senses [] is non-existing, and this [is] at most a clue to that which exists in itself.' This is almost identical to the corresponding statement by Poincaré (and by many metaphysical philosophers and mystics)!

However, what renders Cantor a mystic without any doubt is the following obscure tidbit: he told his good friend, the Swedish mathematician Gösta Mittag-Leffler, that,[44] 'his transfinite numbers had been communicated to him from a *more powerful energy*; that he was only the means by which set theory might be made known.' (emphasis mine). Cantor's claim

---

[43] [Thi], p. 535.
[44] [Dau], p. 290.

that he had a muse, or a *higher source*, who transmitted insightful mathematical ideas to him, is not exceptional. In Chapter 8, we will encounter the Indian mathematician and colorful mystic Srinivasa Ramanujan, who received his lavish formulas directly from a personal goddess.

We point out that Cantor's revelation was communicated to Mittag-Leffler *before* Cantor's first mental crisis in 1884.

This is mentioned in Dauben's biography about Cantor (footnotes 28 and 29 in Dauben's book). To his credit, Dauben does not dismiss this episode as a joke; he acknowledges its relevance to Cantor's life. Yet, he does not go further than treating it as a quirk of an overzealously religious person. We have already reached the point of no return. In the context of the direct scientific and anecdotal evidence of clairvoyance, OBEs, and related phenomena – only a minuscule fragment of which can be shown here – we accept the postulate according to which spirits (Gödel), ghosts (Turing), higher energies (Cantor), muses, or apparitions can be 'real' and can manifest 'objectively'. We will encounter more of this as we go.

The *scientifically inclined* reader might infer that Cantor was at times delusional or suffered from hallucinations. Then again, as both Poincaré and Cantor suggested, the basic problem of our existence that we call 'life' is what is *not* a hallucination. True mystics are virtually unanimous: everything is a kind of hallucination, except possibly the ultimate mystic union with the Supreme Consciousness. In particular, we hallucinate ourselves. There is an interesting anecdote recounted by the psychologist Jack Kornfield. Upon meeting the Tibetan spiritual teacher Kalu Rinpoche, Kornfield asked the Tibetan to describe for him in one sentence the essence of Buddhism. The Tibetan replied:[45] 'I could do it, but you would not believe me and it would take many years to understand what I mean.' The psychologist insisted, and the Tibetan's response was: 'You do not really exist.'

We end this section with a mathematical tidbit. Cantor stated that[46] 'the infinite belongs uniquely to God'. Said Pavel Florenski, the Russian mathematician, priest, and mystic, who practiced name worshipping, achieving altered states of consciousness and mystic journeys through repeating ad nauseam a chant with Jesus' name: 'The set of all sets might be God Himself.' Enter Bertrand Russell with what is now known as Russell's paradox: Suppose the set U of all sets exists. Then it contains itself. So,

---

[45] [Gro], p. 216.
[46] [Dau], p. 123.

some sets contain themselves, some do not. Let Z consist of all sets in U that do not contain themselves. Without going into technical details, we will accept that Z is a well-defined set. Is the set Z a member of Z? It follows immediately from the definition of Z that if Z is a member of Z, then Z is NOT a member of Z, and that if Z is NOT a member of Z then Z is a member of Z. Contradiction! So, the set of all sets does not exist. This caused revisions in some notions in the hierarchy of sets that Cantor had initiated.

6.7 A FEW BRIEF ENTRIES

1. Alan Robert George Owen (1919–2003) was a mathematician and a psi-researcher. He published four books on paranormal phenomena. He conducted a notable experiment in Toronto, 1972: a group of members of the local Society of Psychical Research, led by George Owen and his wife Iris, *conjured up* a ghost named Phillip, who in turn produced various poltergeist acts.

2. Courtney Brown (1952–) is a political scientist, with a specialty in non-linear mathematical modeling. He is proficient in remote viewing (the ability to perceive distant objects or scenes), and he has published three books[47] on that subject.

We offer the following in support of the genuineness of 'remote viewing'. Since the 1970s, various American intelligence agencies conducted clandestine programs on remote viewing. Talented remote viewers included Ingo Swann and Pat Price. During one notable experiment, they were given geographical coordinates of a location, and both of them, independently, described a small island in the Indian Ocean inhabited by a few French-speaking meteorologists. The experiment was blind: the experimenter (Russell Targ; we mentioned him in the section on Martin Gardner) did not know to which location the coordinates applied.

3. Mathematicians as Guinea Pigs

Quoting from Charles Tart (editor): *Altered States of Consciousness*, 1969 (1990 updated edition) page 328:

---

[47] [Bro1], [Bro2].

McCord [H.] and Sherrill[48] reported a provocative experiment with a mathematics professor, a university colleague of McCord's. After assisting the mathematician in entering a hypnotic trance McCord suggested that when awakened he would be given some calculus [!] problems and would be able to do them with high accuracy and at a faster rate than he had ever done such work before. The subject was then roused from his trance, provided with the calculus problems, and asked to solve as many of them as possible in 20 minutes. The subject completed in 20 minutes a task normally would have taken him two hours.

According to the hypnotherapist Dolores Cannon, mathematicians are the hardest to hypnotize. My own experience supports this claim: a professional hypnotist failed to put me under despite my desire to experience it.

---

[48] *American Journal of Clinical Hypnosis* Vol. 4, July 1961, p. 124.

CHAPTER 7

The Many Lives of Bertus Brouwer

Be silent or let thy words be worth more than silence.

PYTHAGORAS

7.1 FIXED POINTS

A cup of coffee theorem: no matter how much one stirs a cup of coffee there will always be a point that stays in the original position.

A cup of coffee

L. E. J. Brouwer

DOI: 10.1201/9781003282198-7

The preceding 'theorem' is, of course, a rather loose statement; after all, coffee is not made of points. A more precise version follows.

A closed ball consists of the points in and on the bounding sphere

A closed disk consists of the points in and on the bounding circle

The fixed-point theorem (three dimensions): for any continuous rearrangement of the points in a closed ball, there must be a point that is fixed (a point that stays at or returns to the original position).

There is also a two-dimensional version of the fixed-point theorem: for any continuous rearrangement of the points in a closed disk, there must be a point that is fixed.

These theorems were discovered and proven by the Dutch mathematician Luitzen Egbertus Jan Brouwer, also known as Bertus Brouwer (1881–1966). In fact, he proved much more: he showed that the corresponding statement is true in every dimension.

The fixed-point theorems fall within the branch of classical mathematics called topology. Through them and the associated theory of modern topology that he founded, Brouwer propelled himself into the elite group of the most prominent mathematicians of the first quarter of the 20th century.

Brouwer also invented *intuitionism*. While topology is within the framework of classical mathematics, intuitionism is not: intuitionism was based on an original (at the time) philosophy of fundamental mathematics, wherein some postulates and axioms of classical mathematics are rejected. For example, intuitionism rejects the claim that every formal statement is either true or false. In a lecture in 1922, he called this principle a superstition. Brouwer's opinion was eventually supported by Gödel's incompleteness theorem mentioned earlier. Another notable statement that intuitionists rejected is (the full form of) the axiom of choice: that one can choose a single element from each non-empty set that belongs to a (possibly infinite) class of sets.

Topology and intuitionism are the two fields of mathematics where Brouwer is remembered the most. In a way he was a kind of mathematical Picasso: proving himself in classical art (topology) before devoting himself to abstract expressionism (intuitionism).

7.2 A SHORT BIOGRAPHY

L. E. J. Brouwer was born in 1881 near Rotterdam, the Netherlands. At 16, he enrolled in the University of Amsterdam to study mathematics. He completed his undergraduate studies in 1900 with excellent grades. He continued with his graduate studies, obtaining a *doctorandus* (MSc) degree in 1903.

At around that time, he started participating in the activities of certain 'new-age' communes. Later, in 1915, he joined the 'Significs' philosophical enterprise founded by the Dutch mystic Frederic van Eeden. Brouwer maintained contacts with mystics or mystically-inclined people throughout almost all of his adult life.

In 1904 Brouwer married Lize de Hall, 11 years older than him, thus instantly becoming the stepfather of Lize's daughter Louise, only 12 years younger than him. Both Lize and Louise will play significant roles in our narrative.

In 1905, provoked by the views of a certain philosophy professor, Brouwer published a pamphlet titled 'Life, Art and Mysticism', which will be one of our main sources of information regarding Brouwer's mysticism. We will discuss it in a separate section.

Brouwer got his doctoral (PhD) title in 1907. His PhD supervisor eliminated three chapters from his originally submitted dissertation; they contained his early thoughts on intuitionism (without using that terminology), intermixed with mystical and 'new-age' proclamations.

In 1913, he became full professor at the University of Amsterdam. Between 1909 and 1913, Brouwer did outstanding research, inaugurating modern topology and proving the fixed-point theorems mentioned above, among other results and ideas. His first significant paper in intuitionistic mathematics was published in 1919.

In the period 1914–1928, he was a member of the editorial board of *Mathematische Annalen* (*Mathematical Annals*), the leading mathematical journal at the time. This was the era when the editors of mathematical journals thoroughly scrutinized the articles submitted for publication and only then distributed the papers for further reviewing. Brouwer was very conscientious in his work, and editorial chores took a lot of his time and energy. He was also rather quarrelsome for a person who professed Buddhist-like beliefs. When he left *Annalen*, he was in conflict with many of the other editors, notably with David Hilbert, perhaps the leading mathematician of the first quarter of the 20th century. All of this took a heavy toll on his health, both mental and physical.

In 1929, his briefcase, containing his scientific diary with his thoughts and ideas from the period of the previous three years, was stolen while he was waiting for a tram in Brussels. This would be a nightmarish event for any mathematician, and, despite Brouwer's remarkable memory, he was not an exception. This incident, problems with his health, as well as the numerous conflicts between Brouwer and other mathematicians, may have all contributed to the subsequent decade-long hiatus in Brouwer's research.

Brouwer lived through both world wars without experiencing any direct personal tragedy: no close relative or close friend perished in these two wars. In both cases, the Netherlands declared neutrality, which was not respected during the Second World War. As a consequence, between 1940 and 1945, there was a scarcity of food, and life was not easy. During these times, Brouwer's family was helped by Max Euwe, one of his relatively few PhD students. Euwe brings up the curious chess theme that threads through the lives of quite a few of our mathematicians: he was the chess champion of the world for two years, between 1935 and 1937.[1]

Brouwer was a passionate traveler, especially during the second half of his life. Naturally, most of his travels were confined to Europe. However, in 1953 he visited Canada and the USA and gave colloquium talks at several universities. In Princeton, he visited Gödel at his home. Gödel, in a letter to his mother, described Brouwer as 'a famous man, already 72 years old, and no longer quite up to date'.[2] The Canadian portion of the trip was organized by the notable geometer Donald Coxeter,[3] who a year earlier had attended Brouwer's talk at the St. Andrews conference, United Kingdom.

---

[1] Euwe was also the president of the world chess federation from 1970 to 1977. In that capacity he visited my native city of Skopje, Macedonia (then in Yugoslavia), where there happened the 24th chess Olympiad. Since it is highly likely that Euwe had played chess with Brouwer, I have a *Brouwer-chess number* of 3.

[2] [Dal2], p. 871.

[3] Coxeter attended a colloquium talk by Brouwer, and I attended a colloquium talk given by Coxeter to the graduate students at the University of Toronto. This gives me the Brouwer-colloquium number of 2.

Brouwer lecturing
during his North
American tour

Though he was well-traveled, to the chagrin of many of his friends, Brouwer never took heed of traffic; he simply ignored it. One evening in 1966, this happened for the last time: he was hit by a car, knocked off into the opposite lane, and rolled over by at least two more cars. He died on the spot.

In his will, he stipulated that the personal correspondence with his wife Lize (who had died seven years earlier) be destroyed, which was duly carried out. He bestowed all of his property on Cor Jongejan; more about her is in the following section.

7.3 BROUWER'S PERSONALITY

Brouwer lived his life as if it was an act in which he was given several mutually independent roles to play. He called these roles personalities or representations, and he took them very seriously. There was Brouwer the mathematician, Brouwer the secretive family man, Brouwer the traveler, Brouwer the flamboyant bon vivant and seducer, Brouwer the humble mystic… However, his case was not of a multiple-personalities syndrome; in the background of all of these aspects of himself there lurked his intelligence, honesty, and zest for living life in full.

Brouwer was an extremely talented mathematician. His discoveries were original and groundbreaking. However, he had an exaggerated sense of justice that made his journey into the world of mathematicians rather exasperating. He had long-standing quarrels with many mathematicians, notably Henri Lebesque and David Hilbert. The issues were often rather banal: who has the priority of a result, or whether a mathematical error is superficial or fundamental. Brouwer was seldom yielding and preferred sacrifices to unprincipled compromises.

His family life was far from being clichéd. To start with, Brouwer's relationship with his 12-year younger stepdaughter Louise was terrible from the get-go and remained such throughout their lives. There was mutual antipathy, amplified by the young Brouwer's obvious unpreparedness to play the role of the father of a teenager. For example, he would not hesitate to slap Louise if she didn't follow his instructions – though one should refrain from using today's moral or social standards to evaluate incidents from a long time ago. Louise, on the other hand, was a slow learner and rather taciturn. When she was 21 and enrolled in a 'domestic science school', Lize and Bertus decided to bring into their household Louise's classmate Cor Jongejan. Cor was an extroverted and fun-loving young woman, the opposite of Louise. The idea was to uplift Louise through companionship with Cor and thus help her in her studies and her life in general. As it turned out, Cor uplifted Bertus instead.

At first, Cor started helping Brouwer with his editorial work: during the same year when Cor appeared, Brouwer became the editor of the *Annals*. By degrees, she became Brouwer's indispensable helper, his official, university-salaried assistant, and companion during his many travels and visits. It became clear that their relationship was far more than a business arrangement. Meanwhile, the role of Lize in Brouwer's life transformed into a mother-like figure. She was a pharmacist and herbalist, and it was she who nursed him back into health when he was sick. Louise, on the other hand, left the house bitter and filled with hate for Brouwer.

Cor to the left, Brouwer, Lize to the right of Brouwer

Women were Brouwer's weakness, and he had many love affairs throughout his life. His flamboyance is no better illustrated than by the following two anecdotes. Once during a discussion between Brouwer, Cor, and two other ladies, all three of them likely Brouwer's lovers, with other people present, he disappeared for a few minutes and reappeared stark naked.[4] The discussion continued as if nothing happened. On another occasion, Brouwer was together with another triplet of women, Cor, Tine, and Gerda. He lay down on the couch and declared:[5] 'I want to be caressed.' Gerda replied: 'Tine, why don't you do it, the way I do it is not good enough anyway.'

Lize died in 1959. Both she and Cor devoted their lives to Brouwer. So it was not a surprise when he proposed to Cor towards the end of their lives. She refused, but remained with him until the end of his life. She died in 1968.

7.4 BROUWER: A TRUE MYSTIC

Recall that the mystic road starts with the empirical ego in the state of being detached, even for a moment, from the mundane din of thoughts and percepts, and ends with a timeless, spaceless realm of the transcendental self, where one experiences blissful unity with the Supreme Consciousness, or God.

Brouwer was a true mystic: he traveled far along the inner mystic roads. This is an unavoidable conclusion from his writings, especially from those produced while he was still a teenager or a young man.

It seems that a mystic experience in the later stages of the life of a person affects the person's philosophy. Almost all of the mathematicians we have encountered so far fall within this category. The same experience, when happening in the early stages of life, affects the psyche of the experiencer. In the latter case, the mysticism becomes a fundamental part of who one is. Brouwer was a young mystic and he remained a young mystic all of his life.

The following is an extract from a 'confession of faith' that 17-year-old Brouwer was required to write in order to join a certain Protestant church; it is a remarkable statement, strongly indicating that Brouwer's mystic travels started early in his life:

> The concept of time, like space, belongs to my representations, whereas my [transcendental] ego is completely separate from these concepts. [] I already mentioned [] my unlimited trust in God, and my conviction of my immortality. [] Among the representation which my God gives me are those that make me at some moments feel intensely his existence, this is then followed

---

[4] [Dal2], p. 719–720.
[5] [Dal2], p. 737

by a strong self-confidence and a joyous courage to live. Each time when this awareness forcefully trust itself upon me, stirring my inner life, I may speak of love for my God. For me such moments of contact do not have the character of prayer, because my wishes and sorrows do not play a role but have disappeared for me.

The quote is from pages 18–19 of the excellent comprehensive biography *Mystic, Geometer, and Intuitionist; The Life of L. E. J. Brouwer* (Volume 1 out of 2) by Dirk van Dalen, our main source in this chapter. Van Dalen, who did not treat Brouwer's mysticism as a joke or as superstition that required cleansing by 'serious' biographers, comments:[6] 'Here one recognizes what traditionally have been called mystic experiences.' Indeed!

The teenage Brouwer's mystic experiences shaped not only his philosophy but also his personality and his life in general.

The main document concerning Brouwer's mysticism is the abovementioned pamphlet 'Life, Art and Mysticism', written by him when he was 23. We quote from Walter P. van Stigt's translation:[7]

This turning-into-oneself requires an effort; it seems that some inertia must be overcome, that your attention is strongly inclined to linger where it is, and that the resistance felt in the move toward the self is much greater than in the move away from it.

If, however, you succeed in overcoming all inertia and proceed, you will find that passions will be silenced, you will feel dead to the old world of perception, of time and space, and all other forms of plurality; and your eyes, no longer blindfolded, will be opened to a scene of joyful quiescence.

It is perfectly clear from this quote that Brouwer had attained mystic union (*Nirvana* in Buddhism, *Samadhi* in ancient Sanskrit texts, *Rapture* in Christianity) at some point before the age of 24!

Here are a few more quotes regarding Brouwer's mystic union, from the same source:

You will see how in this imprisoned world miracles continually break through.

---

[6] [Dal], p. 19.

[7] [Brou].

You will feel endowed with an all-embracing knowledge; as in all emanations you feel the timeless direction, a unison of past, present, and future within yourself.

You will then see how fear and an obsession with saving, born from the illusion of time, and how desire and lust for power, born from the illusion of space, made you attach intrinsic importance to what should only be a fleeting emanation of the self without any reality of its own.

It is remarkable how nonchalantly Brouwer disregarded his own mystical tenets in his daily, mundane life. For example, he made a clear-cut disjunction between the serene mystical world of detached introspection and his simmering wrath against any perceived injustice in his mathematical life.

One more quote from 'Life, Art and Mysticism': 'The life of the individual is illusion …'. Later on, he expounded: 'The system of the heavenly bodies is nothing but a mathematical system, freely built by us, of which people are proud only because it is in this way effective in controlling the phenomena.' He was basically saying that the whole material world is a projected hallucination, or at least that its existence is subjective.[8] As we saw earlier, Poincaré and Cantor had similar views.

Brouwer was a practicing mystic his entire life, and he regularly attended mystic or similar events. He took part in spiritistic seances, and in 1926 he was a participant in the Sufi Congress in Paris. Much later in his life, he attended a lecture by the mystic philosopher Krishnamurti, whom we will encounter again in our story.

7.5 BROUWER THE GHOST?

A long time ago, around 1640, Henry More wrote the following[9] in his treatise *The Immortality of the Soul*:

The examples […] of the appearing of the Ghosts of men after death are so numerous and frequent in all mens [men's] mouths, that it may seem superfluous to particularize in any.

---

[8] Which is true at the level of elementary particles; experiments show that how they manifest depends on the manner we perceive them.

[9] [More], p. 170.

Gödel was wrong when he claimed that 'spirits' [ghosts] manifested in the Middle Ages, and that they[10] 'keep [themselves] in the background today and are not known.' They are as ubiquitous in recent times as they were in the Middle Ages. The difference is in the way the 'modern' society treats this phenomenon today, as compared to the distant past. During the last century or so, the phenomenon of 'ghost' has been ridiculed so much that it has become a caricature of what it used to be, while on the other flank of the war against anything that challenges the materialistic party line, the orthodox medical establishment pathologized 'hallucinations' of any sort into symptoms of schizophrenia or similar.

Ghosts are real; in fact, the thesis that we are as much hallucinations as 'ghosts' is reasonable. The ostensible *reality* of the so-called ghosts is supported by hundreds of *published, veridical* cases, and hundreds of *published* cases of apparitions perceived by more than one person at the same time, virtually excluding the psychological view that they are only projections of our subconsciousness. And, of course, we should not disregard the scientifically established context, a tiny fraction of which we have considered earlier; as Turing said, the phenomenon of 'ghosts' is not far removed from the phenomenon of clairvoyance. We will see more of it in the following chapters; however, the scope of this book does not allow full elaboration.

According to ghost lore, a large subclass of the ghost phenomenon concerns cases of unredeemed 'ghosts' who have not atoned or have not settled the scores with people they knew antemortem. The ghost of Brouwer, the hypothetical Bertus Brouwer (HBB), belonged to this category.

We recall the troublesome relationship between Brouwer and his not-much-younger stepdaughter Louise. Louise hated Brouwer and considered him a fool. Brouwer, at his end of the relationship, seemed to have considerably fueled Louise's resentment: he was not only impatient with her but at times he betrayed his own antipathy towards her. And so it happened that Brouwer-the-ghost, the HBB, haunted Louise for about 14 years!

In 1980 and on the occasion of a conference dedicated to the centenary of Brouwer's birth, planned for 1981, van Dalen wrote a biographical article that appeared in a Dutch weekly. Louise read the article and[11] 'suddenly realized that Brouwer was not a fool after all. As she regularly

---

[10] [Wan2], p. 152.
[11] [Dal], footnote on page 79.

communicated with the spirits of the departed, she noted that Brouwer's spirit had found rest after this'. Louise died the same year.

As we continue our anti-chronological journey, the sources that are available will be more and more scarce and less reliable, a few exceptions notwithstanding.

CHAPTER 8

Srinivasa Ramanujan

Without mysticism man can achieve nothing great.

ANDRÉ GIDE, *THE COUNTERFEITERS*

8.1 AN AMAZING FORMULA

$$\frac{1}{\pi} = \frac{2\sqrt{2}}{9801} \sum_{k=0}^{\infty} \frac{(4k)!(1103 + 26390k)}{(k!)396^{4k}}$$

(The *factorial* ! in the formula is defined by $n! = n(n-1)(n-2)\cdots 2 \cdot 1$. For example, $4! = 4 \cdot 3 \cdot 2 \cdot 1$, which happens to be 24.)

This amazing and difficult-to-prove formula was not a byproduct of a theory. It was delivered by a goddess and received by Srinivasa Ramanujan. At least that was what Ramanujan himself said. Yes, things are becoming outrageous, Horatio.

And, in a way, so is our concept of the reality.

Srinivasa Ramanujan

8.2 RAMANUJAN: A SHORT BIOGRAPHY

Srinivasa Ramanujan was born December 22, 1887, into a Tamil family in Erode, Southern India (today Tamil-Nadu state). It has been said that he was named after the saint Ramanuja (~1100 AD) who had similar astrological signs. He rarely spoke until he was three years old, and his parents

DOI: 10.1201/9781003282198-8

were worried that he was, in the jargon of his time, dumb. As a school-boy, as was the case with the other ingenious people that we have covered so far, he showed his mathematical prowess and received various awards. However, his interest did not extend to other subjects covered in college, and he failed the final (*Fellow of Arts*) exam several times, the last time being in 1907.

In 1900, when he was 12, he got a book in trigonometry (written by S. L. Loney), and when he was 16, he borrowed a library copy of G. S. Carr's book which listed, mostly without proof, some 5,000 theorems. These two, especially the latter, widened his mathematical horizons and shaped his interests. They also stimulated his extraordinary creativity: when he was 15, he developed his own method for solving quadric equations, and he tried unsuccessfully to do the same for quintic equations (without know-ing that Niels Henrik Abel showed in 1824 that that was not possible[1]). In any case, the irresistible spell of mathematics got hold of him and stayed with him until the very last days of his life.

And so, mathematics was what he was mostly doing during the next three to four years following his last attempt in 1907 to become a college graduate: he was entranced by his mathematical theorems while living in abject poverty. In 1909, there happened a little side story in his life: he got married. His mother found the pretty Janaki, a bride for him. The differ-ence in age between him and Janaki was about the same as the difference in age between Brouwer and Lize; except for the sign: Janaki was ten when they got married. She stayed with her parents until she was 14.

Between 1910 and 1912, Ramanujan tutored students in Madras (now Chennai), then managed to secure a modest clerical position. More impor-tantly, he established contacts with some Indian mathematicians who rec-ognized his indisputable mathematical talents. Eventually, he secured an exceptional research position at Madras University, and his financial situ-ation improved markedly. It helped that he was from the Brahmin caste, which had the highest status among the four social classes of India.

In 1913, his notebooks were already loaded with hundreds of highly original theorems. At about that time, he was advised to contact some prominent British mathematicians and share the notes with them. The first two he had contacted by mail to England returned the packages with-out comments. Apparently, they judged him to be a crank, which was not

---

[1] For example, $x^5 - 4x^3 - 2 = 0$ is a quintic equation (5 is the highest power) that is not solvable by radicals (roots).

surprising since no proofs were supplied, and since some of the formulas looked ridiculous at first sight. Consider, for example, the following: $1 + 2 + 3 + ... + \infty = -\dfrac{1}{12}$. Ramanujan himself was aware of the potential for misunderstanding and emphasized that this result is true in *his* theory, but that apparently did not help much.

The third prominent British mathematician that Ramanujan contacted by mail was the number theorist Godfrey Harold Hardy. Hardy recognized the originality and the brilliance of the results and responded, to some extent cautiously, but also encouragingly.

Eventually, Ramanujan ended up in Cambridge, England, and stayed there between the years of 1914 and 1919. He was awarded a stipend and devoted most of his time to mathematics, working with Hardy and occasionally with Hardy's collaborator John Edensor Littlewood (1885–1977). He started well and published nine papers in 1915.

In 1917, Ramanujan got sick and was sent to a hospital. Initially, he was diagnosed as suffering from a gastric ulcer; later it was confirmed that he had tuberculosis. Ramanujan had a very tough time in the cold English hospitals. It did not help that he did not have faith in conventional medicine to start with.

Despite relatively frequent contact with Hardy – though there were periods when they did not see each other for months – Ramanujan was often very lonely. He missed his native country: his family, the exuberant Indian social life, the warm climate, and the food. His mental and physical health was declining and reached a low point at the beginning of 1918 when, in a state of despair, he attempted suicide by throwing himself in front of an approaching train in a subway station. His life was saved by a vigilant guard who promptly applied the brakes and brought the train to a screeching stop a few feet from the lying Ramanujan.

In May of the same year, Ramanujan was elected Fellow of the Royal Society (FRS), the most prestigious scientific honor in Great Britain. Hardy, who was himself an FRS, helped considerably through spirited lobbying on Ramanujan's behalf.

In the spring of 1919, Ramanujan returned to India. By that time his health had deteriorated significantly, but he continued his mathematical work. He died in 1920, aged 32.

8.3 RAMANUJAN AND HARDY

Ramanujan was a strict follower of Hinduism and was zealously religious. For example, the main reason for his initial ambivalence with regard to

the prospect of traveling to foreign lands was because, according to orthodox Hinduism, such travels would affect his Brahmin purity. Ultimately,
he decided to travel, but only after he had received explicit permission, no
less than from a goddess! This brings us to the main topic of our story in
this chapter: Ramanujan's mysticism.

Mysticism has never been far from Hinduism, and, within this scope,
it was never regarded as heresy. In fact, it has been incorporated within
Hinduism in many ways. Hinduism, being mostly devoid of clerical
orders, has always been more a religion of experiencers rather than of
followers, and deliberate mystic travels and contacts, sometimes all the
way to the mystic union (*Samadhi*), were ways to express adherence to its
principles in practical terms and were sought by the initiated as well as by
ordinary people.

Ramanujan and his maternal grandmother were both mystic travelers.
We start with his grandmother, and quote from Robert Kanigel's *The Man
Who Knew Infinity*, one of the very few biographies that do not cleanse
Ramanujan's life from his mysticism.[2]

> Ramanujan's maternal grandmother, Rangammal, was a devo-
> tee of [the goddess] Namagiri [their family's deity] and was
> said to enter a trance to speak to her. [] Another time, [] before
> Ramanujan's birth, Namagiri revealed to her that the goddess
> would one day speak through her daughter's son.

And so, she did. Ramanujan would invoke Namagiri's name, and she
would offer counsel, advice, and revelations. That was exactly what happened when he needed guidance regarding the proposed travel to England.
Ramanujan, accompanied by a friend, visited the shrine of Namagiri in
the town of Namakkal, and there, after chanting Namagiri's name for
three days and two nights, he had a vision during which Namagiri manifested and gave him an *adesh* (command) to bypass the injunction against
foreign travel.[3] From the point of view of orthodox psychology, this was a
case of delusion or self-suggestion by someone who dearly wanted to continue along his mathematical path. We do not accept that.

Ramanujan's mysticism extended beyond Hinduism. He believed in
astrology, as well as palmistry: he predicted from the lines of his palms
that he would not live past 35. In England he was interested in the work

---

[2] [Kan], p. 36.
[3] [Kan], p. 188.

of the parapsychology researchers: we know that he studied the work of the physicist Oliver Lodge, whom we mentioned in Sections 6.3 and 6.4.1.

Hardy was far from being religious; he was a staunch atheist. He had leftist political views and displayed a photograph of Lenin on the mantelpiece in his house. According to his main collaborator Littlewood, he was also a non-practicing homosexual.

Their social upbringing was very different too. Indian social life was abundant: it was normal for strangers to come into Ramanujan's house, and friendly personal contacts were often virtually instantaneous. It was not unusual for new acquaintances to exchange private details of their lives after a few minutes of conversation. Hardy, on the other hand, was the epitome of English reticence and avoided intimate personal contact. He did not know much about the human side of Ramanujan; Ramanujan-the-mathematician was sufficient to him. Indeed, Hardy once wrote that[4] 'Ramanujan was no mystic [] and religion played no important part in his life.' He could not have been more off the mark. Even fellow Englishmen regarded Hardy as being overly reserved. Turing met him in Princeton in 1936 and found him[5] 'very standoffish or possibly shy. I met him the day I arrived [] and he didn't say a word to me.'

It is strange how Hardy and Ramanujan's mathematical incongruencies were mutually complementary, joining into a kind of mathematical symbiosis. Hardy was a perfectionist who measured his words and paid attention to the minute details of his proofs. During his early mathematical career, and to a lesser degree later, Ramanujan couldn't care less about proofs; for him, the truth of a formula was a consequence of its metaphysical significance, measured in terms of beauty, transcendence, and universality. He once said:[6] 'An equation for me had no meaning unless it expresses a thought of God.'

Hardy was aware of Ramanujan's mathematical weaknesses, and he tried to help him in repairing the gaps in his mathematical education and attitude. However, he thoughtfully decided not to exert too much pressure, being afraid that insisting on proofs and justifications might negatively affect Ramanujan's mathematical intuition.

Hardy, in his own way, was a good and compassionate person. For example, his lobbying for Ramanujan's admittance into the Royal Society

---

[4] [Kan], p. 283.
[5] [Kan], p. 278.
[6] [Kan], p. 67.

was to a large extent motivated by his desire to uplift Ramanujan and help him recover from his mental and physical crisis. Hardy himself was a powerful mathematician; regardless, he was rather modest: on a scale from 0 to 100 he gave himself 20, he gave Littlewood 30, Hilbert was ranked 80, and Ramanujan 100.

Godfrey Harold Hardy

Hardy also tried to commit suicide. He swallowed barbiturates, but, in a morbidly ironic twist, he took too much and vomited. This happened in 1947. He was already 70 years old, sick, and worn out. He died in December of that year.

Twelve years earlier, in 1935, on the occasion of the 300th anniversary of the founding of Harvard, there was a ceremony honoring 62 distinguished scientists. Hardy was one of them. Another one in the procession was the outstanding psychoanalyst and psychiatrist Carl Jung. In 1915, in order to explain some veridical hallucinations that were not related to the personal histories of the experiencers, Jung postulated the existence of (what he called) collective unconscious. Collective unconscious is a cache of pre-existent primordial images and ideas, or archetypes, encompassing the soul of humanity, individually accessible through dreams, visions, and other inner paths within the subconsciousness. In Jungian terms, Ramanujan found his theorems in the collective unconscious.

Later in his life, Jung came to believe in the hypothesis that consciousness, in particular human consciousness, transcends matter. Here is a quote from *Face to Face with Professor Jung*:[7] 'The psyche, in part at least, is not dependent on this confinement [body, time, space].' Perhaps including the matter-less, discarnate consciousnesses within the pool of artifacts would then be a consistent expansion of his theory of 'collective unconscious'. This brings us to the next section, where we postulate the main thesis of this chapter.

8.4 MATHEMATICS, DREAMS, HALLUCINATIONS: RAMANUJAN AND NAMAGIRI

As we indicated earlier, there is ample scientific, statistical evidence, as well as a vast amount of anecdotal evidence, establishing that clairvoyance is genuine. There are many experiments under strictly controlled parameters, many more experiments where some of the parameters are out of the control of the experimenter, and a huge number of anecdotal cases, many with veridical elements, altogether tracing the path between clairvoyance and OBE. Once we accept the genuineness of OBE and the related phenomena, we open the floodgates wide open, and through them flows the thesis that consciousness, human or other, is not bound to a corporeal conveyor; we have discussed this issue in Chapter 6. From this claim, it follows that postulating the existence, within our 'reality', of disembodied, sentient entities, or 'ghosts', becomes only a matter of accepting the extremely well-substantiated thesis that disembodied consciousnesses can sometimes enter our spectrum of perception.

Back to Ramanujan's mathematics, here is a cute-looking formula from his early mathematical adventures:

$$3 = \sqrt{1+2\sqrt{1+3\sqrt{1+4\sqrt{1+...+n\sqrt{1+(n-1)(n+1)}}}}} \ , \ n=1,2,3,....$$ It is an example of a clever formula that is relatively easy to come by and, therefore, easy to justify: repeatedly using the simple identity $n+1=\sqrt{1+n(n+2)}$, we get $3=\sqrt{1+2\cdot4}=\sqrt{1+2\sqrt{1+3\cdot5}}=\sqrt{1+2\sqrt{1+3\sqrt{1+4\cdot6}}}=...$, and we see its origin and a justification at the same time.

Now, consider the following theorem from the last page of Ramanujan's first letter to Hardy:

[7] www.youtube.com/watch?v=hD-W-1z_qco.

If $u = \cfrac{x}{1+\cfrac{x^5}{1+\cfrac{x^{10}}{1+\cfrac{x^{15}}{1+\ldots}}}}$ and $v = \cfrac{\sqrt[5]{x}}{1+\cfrac{x}{1+\cfrac{x^2}{1+\cfrac{x^3}{1+\ldots}}}}$ then $v^5 = u\dfrac{1-2u+4u^2-3u^3+u^4}{1+3u+4u^2+2u^3+u^4}$

Ramanujan seldom or never supplied any proofs, and this theorem is such an example. A ten-page proof was eventually supplied by L. J. Rogers in 1921, some *eight years* later. Hardy discussed this and other of Ramanujan's theorems with Littlewood and concluded: 'They must be true because, if they were not true, no one would have the imagination to invent them.' Indeed!

If this is not enough to convince you that something extraordinary was happening with Ramanujan's theorems, then consider the following truly mind-boggling result:

$$1+\frac{1}{1\cdot3}+\frac{1}{1\cdot3\cdot5}+\frac{1}{1\cdot3\cdot5\cdot7}+\ldots+\cfrac{1}{1+\cfrac{1}{1+\cfrac{2}{1+\cfrac{3}{1+\cfrac{4}{1+\ldots}}}}}=\sqrt{\frac{\pi e}{2}}$$

I guess the most natural reaction upon seeing such a marvel is to regress into a child's mentality and say 'wow!'. So, I say 'Wow! What an exquisite beauty!' And how utterly mystifying: how on earth does the product of π and e enter the dance played by the integers on the left-hand side of the identity? And why would one expect the *sum* of the series and the continuous fraction to be $\sqrt{\dfrac{\pi e}{2}}$??

The extraordinariness of this theorem is compounded manifold by the manner in which it had been obtained: in Ramanujan's notebooks this formula was not derived from a theory; it is a standalone statement, extracted straight from the blue sky, so to speak. Or, perhaps it was a gift to Ramanujan from the discarnate entity identified as goddess Namagiri? Assuming sentient discarnate entities exist, one is tempted to postulate that this is the *only* way a human can get hold of such a transcendent isolated result! We postulate exactly that!

First of all, Ramanujan himself would have likely concurred: he was humble and honest, and he attributed at least some of his theorems to the goddess Namagiri. He told[8] his friend T. K. Rajagopolan that after seeing drops of blood, 'scrolls containing the most complicated mathematics used to unfold *before his eyes*' [emphasis mine]. It seems that we are dealing here with visions or 'hallucinations', rather than dreams. Unlike dreams, hallucinations are perceived with open eyes.

Most of the biographers, including van Dalen, and most of the experiencers as well, tend to *dreamatize* hallucinations since they carry pathological connotation, while dreams are normal. Dreams have always been, in a way, good guardians against any nonmaterialistic interpretations of mystic episodes. Most people are not aware that nonpathological hallucinations of healthy people are also ubiquitous: as we have mentioned earlier, numerous polls show that the majority of people experience at least one hallucination during their lifetime.

However, it is not important to us whether Ramanujan perceived his mathematical scrolls in dreams or through hallucinations. We are interested in the reasonable thesis that discarnate entities exist and that they can communicate with us, humans. For the latter to happen usually one needs to be in an altered state of consciousness; dreaming and hallucinating are the most common such states.

Such supernatural contacts have been recorded and published thousands of times. We will consider only a few instances of experiences similar to Ramanujan's.

The following was reported by Romaine Newbold in *The Proceedings of the Society for Psychical Research*, vol. XII, pages 11–20, year 1897. The title of the article is *Subconscious Reasoning*. William Lamberton, a professor of Latin and Greek, befriended the math professor in a school and started studying *descriptive geometry*. He tried to solve the following problem: find the locus of the foots of the perpendiculars from a focus of an ellipse to all of the tangents to the ellipse. He could not make much progress and decided to stop thinking about the problem for a while. Then, one night…

> … on opening my eyes, […] I saw projected upon this blackboard surface [on one of the walls of his sleeping room] a complete figure, containing not only the lines given in the problem, but also a number of auxiliary lines, and just such lines as without further thought solved the problem at once.

---

8 [Kan], p. 281.

A solution through a hypnogogic hallucination! In orthodox terms, this would be a case of an external projection of the subconscious. In Jungian terms, the solution was found in the collective unconscious. In mystic terms, someone out of this world intervened.

Musicians and mathematicians seem to be the most susceptible to such interventions.

Robert Schumann (1810–1856), the great German romantic composer, was a voice hearer, and, according to his wife, he received his music through auditory hallucinations. Clara (his wife) wrote[9] that 'he heard entire pieces from the beginning to end, as if played by a full orchestra.'

We also have the curious case of the Italian composer and violinist Giuseppe Tartini, who had a 'dream' one night in 1765 during which he gave his violin to a devil. The devil then proceeded to play [quoting Tartini][10] 'on a level I had never before conceived was possible.' Tartini was angry at himself for not remembering most of the music, but from what he did he wrote the sonata 'Devil's Trill'.[11]

The unearthly bearers of offerings, when they manifest at all, come in many forms, depending on the background of the experiencers. They are perceived as goddesses, beings of light, angels, or simply ghosts.

Academic research on 'ghosts' is dangerous for one's career. So, it is not surprising that it is virtually non-existent. Yet, there are some timid attempts. We mentioned earlier Gary E. Schwartz, an American psychologist, author, parapsychologist, and professor at the University of Arizona and the director of its Laboratory for Advances in Consciousness and Health. Once he *saw*, several times, a shining being in the form of a very tall, blond, long-haired beautiful woman, whom he called Sophie. That motivated him to devise an experiment: someone would simply ask Sophie to enter a dark chamber monitored by a very sensitive (and expensive) photon-catching camera. Then they would analyze the images before, during, and after the presumed Sophie's visit. Two things happened. First:[12] 'We observed a faint angel-like shaped being, with its head tilted toward the upper left-hand corner, and its legs pointing in the lower right-hand corner.' Secondly, they analyzed the images using *Fast Fourier Transformation*

---

[9] [McC], p. 338.

[10] [Ric], p. 153; also in www.youtube.com/watch?v=QiEgOU18oE0.

[11] The story of the French writer Guy de Maupassant is too outrageous for this stage of our narrative. So, I am relegating a synopsis to this footnote. In 1889, Maupassant hallucinated his double (an apparition that looked exactly like himself), and the double dictated the original short story 'Le Horla'.

[12] [Schw], p. 200

(FFT), a method that splits a complicated wave-like function (sound, light, etc.) into simple waves (or Fourier series). The FFT analysis was also positive, showing a clear difference between the Sophie image and the images of the dark chamber before and after the presumed presence of the angel in the dark chamber. Schwartz and his assistant then repeated the test with a change: this time the assistant meditated and invited the *Divine Light* to enter the chamber at a certain period of time. The results were the same: 'FFT ripples occurred [only] during the presence-of-spirit period.' The pattern was the most complex precisely in the case of the presence of the *Divine Light*.

Back to our mathematicians: receiving mathematical insight in dreams or *dreamatized* hallucinations is a fairly common occurrence, and the cases we have encountered so far (Poincaré, Cantor, Ramanujan) seem to be the tip of the iceberg. Interpreting such events as a delivery of information from one realm of existence to another is, on the other hand, uncommon. This is due less to the outrageousness of this (nonmaterialistic) proposition, and more to the degree we have been brainwashed into devotional materialism by the controllers of the 'scientific' narrative. In our narrative, we will stubbornly keep going in the metaphysical direction, away from materialism.

Is it too wild to propose that a 'creative concentration', when a person is deeply immersed in thinking, particularly in mathematical searching, is a trance-like state, a mediumship of a sort, when one accesses the Platonic world of ideas, which presumably exist out of one's mind? If so, then the degree of 'creativity' would be equivalent to the degree of sensitivity: the better 'necromancer' one is, the more intelligent he comes across. The freedom of thought then becomes the same as the ability to access the external, subliminal, and much deeper world of abstract reality, of which we are but a minute part.

We claim that mathematical insights, truly new results, and ideas, or eureka moments, are often cases of channeling: transmitting information from the realms inhabited by pure consciousness into our sensual material world.[13] The greatest mathematicians often combine ingenuity with channeling. This was the case with Ramanujan, and, surprisingly, this may have been the case with the 'king of the mathematicians', Carl Friedrich

---

[13] These claims should be taken tentatively because we don't even know what matter is, or in which sense it exists.

Gauss, who once wrote upon proving a theorem that it was done,[14] 'not by dint of painful effort but so to speak by the grace of God.'

And no, I am not diminishing the achievements of our great mathematicians.

In order to access the metaphysical realms and receive ingenious mathematical gifts, one has to be ingenious to start with. Otherwise, you will get no more than a solution of an exercise about ellipses. The 19th-century psi-researcher Alexander Aksakov recounts the following short story,[15] supporting our arguments in a comical way. His acquaintance I. I. Mousin-Pushkin, after attending a mediumistic séance and being impressed by what he experienced, decided to see if he had some medium-istic potential by trying to invoke spirits. The attempt was successful, for he got in touch (via raps) with his deceased mother. The story ended in an anti-climax: she flatly told him that such contacts were not for him, and she never manifested again.

More about mediums in the next chapter.

---

[14] [Kan], p. 287.
[15] [Aks], p. 323.

CHAPTER **9**

Augustus De Morgan and His Wife Sophia

A goods scientist has freed himself of concepts

And keeps his mind open to what is.

LAO TZU, 6TH CENTURY BC

9.1 DE MORGAN LAWS

(1) $(A \cup B)^c = A^c \cap B^c$

(2) $(A \cap B)^c = A^c \cup B^c$

(3) $\neg(p \vee q) = \neg p \wedge \neg q$

(4) $\neg(p \wedge q) = \neg p \vee \neg q$

These are the simplest instances of De Morgan laws: (1) and (2) are basic set theory; (1) is the statement that the complement of the union of A and B is the intersection of the complements of A and B, and (2) is the statement that the complement of the intersection of A and B is the union of the complements of A and B. (3) and (4) are essentially the same as (1) and (2), respectively, only stated in the language of logic; in (3) we say that the negation of the statement p or q (where p and q are statements) is the negation of p and negation of q.

DOI: 10.1201/9781003282198-9

And this is Augustus De Morgan, one of the founders of symbolic (mathematical) logic.

Augustus De Morgan

9.2 AUGUSTUS DE MORGAN: A SHORT BIOGRAPHY

Augustus De Morgan was born in 1806 in the Indian city of Madurai, some 200 kilometers south of Ramanujan's birthplace. As a baby, he was affected by a disease called 'the sore eye of India' and lost one eye. He had six siblings, and three of them died as children. His father was a lieutenant colonel and held different posts at several stations in India. When Augustus was seven months old, the De Morgan family moved to England; his father traveled to India several times and died on his way back in 1816.

He went through upper-middle-class schools and then college in Cambridge. His mother's wish was that Augustus became an Evangelical clergyman, and initially, he intended to comply. However, the pull of mathematics was too strong, and he eventually followed his natural inclination.

When he was 21 he met Sophia Frend, who was then 19. She described him at the time of their first meeting as a funny young man (who looked older than his age), loving fairy tales and ghost stories.[1] Ten years later, she would become his wife. They would eventually have seven children.

---

[1] [DeM2], p. 20.

Sophia is important for us: she wrote the books[2] *From Matter to Spirit* and *De Morgan, Memoir*, which are our main sources for this chapter. The first edition of *From Matter to Spirit* came with anonymous authors: Augustus, who wrote the preface, was A. B., and Sophia was C. D.;[3] apparently already at their time, there was a social stigma associated with those who took such phenomena seriously. The preface itself was a whopping 45-page essay, sprinkled liberally with untranslated quotes in French, Latin, and Greek, and apparently aimed at an educated readership. The biography *De Morgan, Memoir* was written after Augustus died; this section is based on it. It is interesting to note that in it, Sophia refers to Augustus as Mr. De Morgan!

Augustus was not yet 22 when he was elected to the chair of mathematics in the newly established University College in London, called at first the London University.[4] Unlike many other mathematicians, he enjoyed lecturing, which was quite fortunate because his teaching load as a full professor was enormous, compared with today's corresponding teaching loads: he regularly delivered four courses each of three hours per week, and he had occasional evening classes on top of that.

His most significant mathematical contribution, as we have mentioned, was the mathematical formalization of basic logic. He was also the first who formalized mathematical induction.

Augustus wrote several books, notably *Formal Logic* (1847) and *Budget of Paradoxes* (1872, republished 2007). The latter is a collection of short notes on mostly popular mathematical subjects; it was a well-read book at the time.

Augustus resigned his professorship in 1866 in protest against the decision of the Council of University College to deny the appointment to the chair of mental philosophy (philosophy of mind) of a certain Unitarian minister, even though he was the most qualified candidate; some members of the Council objected to the theist philosophy of Unitarianism.

His second son George died in October 1867. George was also a mathematician, and his father was very fond of him. At that time Augustus was already frail, and after George's passing his health declined rapidly.

---

[2] [DeM] and [DeM2].

[3] Had they written the book today, after a century and a half of steady social *progress*, they would have probably also used pseudonyms (for the stigma is still present), but one would expect Elizabeth to claim A. B., with Augustus settling on C. D.

[4] [DeM2], p. 24.

In August 1870, his daughter Christina died. Augustus lost his will to live and died on March 18, 1871.

9.3 AUGUSTUS DE MORGAN: A RATIONAL MYSTIC

Another mathematician credited with the early development of formal logic was George Boole. Boole held an appointment as the mathematical chair in Cork, Ireland. Augustus De Morgan and George Boole met several times, and, according to Sophia, their meetings were a real enjoyment for both. Sophia also tells us[5] that they both shared the belief that 'every system which rejected the existence of God as a constantly sustaining cause of all mental as well as physical phenomena, was like a consideration of the nature and growth of a tree without reference to the root.' So, the often-encountered claim that Augustus was an atheist is dubious.

What is certain is that he was open-minded and ready to accept the consequences of his intellectual or sensory experience, in whatever direction it led him. This already made him a kind of proto-rational mystic. Exposure to certain phenomena was all that was needed to eliminate 'proto'. Enter his wife, Sophia.

Sometime in 1853, Sophia, with a group of friends, decided to attend a séance of the American *medium* Mrs. Hayden, in whose presence discarnate consciousnesses manifested. The *sitters* (those who attended the séance) and their names were unknown to Mrs. Hayden. They all sat around a table and waited patiently for more than 15 minutes when throbbing sounds began emanating from the center of the table. Eventually, a *spirit* came through and indicated (through raps on the table) his/her/its wish to speak to one of the speakers. To Sophia's surprise, the raps were heard when Mrs. Hayden pointed to her. Sophia was then instructed to point to the letters of the alphabet, and the words were slowly spelled out through raps. Sophia wrote:[6] 'However, to my astonishment, the not so common name of a dear relation who had left this world seventeen years before, and whose surname was that of my father's, not my husband's family, was spelt.' The details of the séance are not important to us. Sophia attended one more séance with Mrs. Hayden, then invited her for a visit to her home.

What happened in De Morgan's house was described by Augustus in the introduction to Sophia's book *From Matter to Spirit*, as well as in a

---

[5] [DeM2], p. 168.
[6] [DeM], p. 13.

letter to Augustus's old friend William Heald (cited in Sophia's *Augustus De Morgan, Memoirs*). We quote from Augustus's letter to Heald:[7]

To Rev. W. Heald. July 1853

I remember giving you my experience in regard to clairvoyance. I will now tell you some of my experience in reference to table-turning, spirit-rapping, and so on. Mrs. Hayden, the American medium, came to my house, and we had a sitting of more than two hours. She had not been there many minutes before some slight ticking raps were heard in the table apparently. The raps answered by the alphabet (pointing to the letters on a card), one after the other (a rap or two coming at the letter), to the name of a sister of my wife, who died seventeen years ago. After some questioning, she (I speak the spirit hypotheses, though I have no theory on the subject) was asked whether I might ask a question. 'Yes,' affirmative rap. I said, 'May I ask it mentally?' 'Yes.' 'May Mrs. Hayden hold up both her hands while I do it?' 'Yes.' Mrs. H. did so, and *in my mind* [original emphasis], without speaking, I put a question, and suggested that the answer should be in one word, which I thought of. I then took the card, and got that word letter by letter CHESS. The question was whether she remembered a letter she once wrote to me, and what was the subject.

(In the first sentence of the quote, De Morgan refers to the phenomenon of (what is now called) distant viewing (Section 6.7), manifested by a hypnotized girl a few years earlier.)

We note and then let go of the curious appearance of the chess motif in our stories, and we pay attention to the phenomenon at hand. Assuming there was no cheating, the miracle of *necromancy* (communication with the *dead*), facilitated by the mediumship of Mrs. Hayden, is the most obvious interpretation. If one accepts the proposition that consciousness survives the dissolution of the body, or death, then this is not a miracle at all. As we mentioned earlier, contact with the disembodied entity is all that is needed to declare the events that transpired in De Morgan's house as merely a kind of corollary of the thesis that disembodied entities exist.

---

[7] [DeM2], p. 221..

Aside from the frivolous 'this must have been cheating' argument, the most acceptable 'normal' explanation would be based on telepathy: clearly, someone was accessing Augustus's thoughts when stating the questions in his mind. Telepathy is a much smaller miracle, and it doesn't require a profound paradigm shift; it is hoped that it might be on the way to linearly expanding orthodox science. However, that does not explain the *miraculous* raps emanating from the middle of the table, while, as we are told, Mrs. Hayden was engaged in reading *Uncle Tom's Cabin*!

We digress for a moment: According to H. B. Stowe, the author of Uncle Tom's Cabin, she did not write the book, but rather it was given to her; just like Ramanujan's formulas, the content of the book 'passed before her.'[8] There is a hidden, subliminal world – in fact, there are many of them – that do not fit the narrow confines of the reality sanctioned by dogmatic scientifism.

Back to the main line of our narrative: the next experiment of Augustus with Mrs. Hayden, or with the manifesting *spirits*, virtually excludes telepathy as the sole explanation.

During the impromptu séance, the spirit of Augustus' father manifested subsequently, and he requested that the manifesting spirit spell the first letters of some phrase that Augustus had in mind that only he and his father knew about. The first three letters, spelled out by raps, were not what Augustus expected, thus excluding telepathy. However, the subsequent letters were correct. Only afterwards Augustus realized that the first three letters were the initials of his father's name.

Augustus and Sophia attended many other séances, and both were convinced of the authenticity of the manifesting phenomena. However, he did not write anything else about such events. Sophia, on the other hand, wrote extensively about her experiences in *From Matter to Spirit* (and in one more book). We will mention Sophia again in the next chapter.

In 1871, Augustus's life ended with an ostensibly paranormal event. We quote from Sophia's *Memoirs*, page 387:

> During the last two days of his life there were indications of his passing through the experience which he had himself considered worthy of investigation and of record. He seemed to recognise all those of his family whom he had lost: his three children, his mother and sister, whom he greeted, naming them in the reverse

---

[8] Source: Nandor Fodor, *Encyclopedia of Psychic Science*, page 22.

order to that in which they left this world. No one seeing him at that moment could doubt that what he seemed to perceive, was, to him at least, visible and real.

Was this a deathbed hallucination caused by a clouded brain? Perhaps. And then again, perhaps not! Suppression of normal perception by the barely functioning brain may have opened a window for other types of perception. In the context of the claim that fully functioning consciousness survives 'death', this is a reasonable postulate. And, of course, there are studies directly supporting this thesis. We will pay brief attention to one such study.

During the 1950s, Karlis Osis conducted a pilot survey:[9] he sent an 11-question questionnaire to 5,000 physicians and 5,000 nurses in India and the USA pertaining to their observations of deathbed experiences by their patients. He was particularly interested in the emotions and experiences of the patients who were fully conscious just before they died.

It turned out that about 10% of the dying people were fully conscious during the last hour before passing over. A significant proportion of these people, 884 cases, or around 40%, reported hallucinations of various personalities, the majority of these (52%) were images of dead persons. Interestingly, in four cases, the apparitions were *perceived collectively* by the patient and by other people. Four *miracles*!

9.4 MEDIUMS

True mediums are people gifted with special extrasensory perception: they can establish rapport between people of the 'ordinary' world and sentient beings residing in realms not separated from us by space but by other parameters (frequency?) and by degrees of awareness. Once the contact is established, these sentient beings manifest in various ways, from the simplest (the medium transmits information) to the most outrageous (full materialization; we will eventually go there too!)

True mediums have been scrutinized and tested a myriad of times under strictly controlled conditions and by reputable scientists and researchers. We will only mention the corresponding investigation by one of the greatest physicists/chemists of the 19th century, Sir William Crookes, who discovered the chemical element thallium. In July of 1870, after being directly influenced by Sophia's and Augustus's writings, he decided to investigate

---

[9] [Osi]; also in [Lor], p. 224.

mediums himself and the alleged phenomena manifesting in their presence. His announcement to that effect was greeted with wide approval by the public, the majority of whom expected a thorough debunking. Crookes wrote:[10] 'The increased employment of scientific methods will produce a race of observers who will drive the worthless residuum of Spiritualism hence into the unknown limbo of magic and necromancy.' In December of the same year, after scrutinizing a few mediums, he wrote[11] something very different in his diary: 'May He [God] also allow us to continue to receive Spiritual communications from my brother who passed over the boundary when in a ship at sea more than three year ago.'

Crookes continued with his investigations for many years, authenticating some of the most outrageous cases of mediumship. All of that is outside the scope of this book. Here, we prefer scientifically controlled investigations that can be statistically evaluated.

Gary E. Schwartz is a professor at the University of Arizona and the director of the Laboratory for Advances in Consciousness and Health at the same university. In 2002, he published the book *The Afterlife Experiment*[12] where in the introduction he says the following: 'This book presents a scientific possibility that [eternal life] has been proved and is real.' He based this claim on the experiments he had performed with some prominent mediums.

The mediums that were employed in his research were John Edward, Suzanne Northrop, George Anderson, Anne Gehman, and Laurie Campbell. In one type of experiment, sitters were quiet the whole time while mediums tried to get some reading on a sitter's life or deceased relatives. In one case, all five mediums correctly stated that the sitter's son had died (nobody mentioned any deceased daughter), and all five of them correctly stated the initial (M) of the deceased son. The odds that this happened by chance are smaller than one out of ten million (there are 9,765,625 ways to choose five letters in the English alphabet). Explaining this away by means of models of telepathy susceptible to orthodox science doesn't work in all cases. For example, in some instances the sitter and the medium, unknown to each other, were in silent contact by phone, long distance from each other.

---

[10] [Smy], p. 154.
[11] Ibid, p. 155.
[12] [Schwl].

Completely excluding telepathy as an explanation of how mediums operate requires the transmission of information not known to any living person, which is subsequently verified. This has been done many times.[13] Coded messages have also accomplished this, as was (allegedly) done in Houdini's case, mentioned in Section 6.4.

We are aware that acceptance of the genuineness of the phenomena described in this chapter requires a distant departure from the common, materialistic scientific theories of reality. It is not as much a matter of a paradigm shift as much as it is a huge paradigm expansion: the theories that allow for the existence of (post-mortem) pure consciousnesses do not necessarily contradict the basic materialistic axioms. On the other hand, the theories stemming from materialistic axioms are too narrow to account for the *outrageous* phenomena that we have, and will, describe, and whose authenticity was believed strongly by the mathematicians we cover in our narrative. Huddling in the warmth of the totalitarian mind-set of the majority and then high-handedly scoffing at such phenomena would not be conducive to their proper understanding.

---

[13] See, for example, [Bon].

CHAPTER 10

Leibniz and Newton

Death is an avoidable accident.

'Emil' in Spalding's *Life and Teaching of the Masters of the Far East*, Vol. 1

Isaac Newton (1642–1726) and Gottfried Wilhelm Leibniz (1646–1716) shared about the same historical time, as well as many traits of their scientific and personal lives. It is only fitting that they share the same chapter in our narrative.

Isaac Newton ...   ...and his 'scientific twin' Gottfried Leibniz

DOI: 10.1201/9781003282198-10

10.1 A SAMPLE OF NEWTON'S MATHEMATICS

(1) Newton's law of universal gravitation is $F = G\dfrac{m_1 m_2}{r^2}$, where F is the gravitational force, (the big) G is the gravitational constant (approximately equal to $6.674 \times 10^{-11}\ m^3\ kg^{-1}\ s^{-2}$), m_1 and m_2 are the masses of two point-like bodies, and r is their distance.

(2) Newton's second law of motion (in modern notation) is $\vec{F} = m\dfrac{d^2 \vec{s}}{dt^2}$, where F is the force, m is the mass, $\vec{s}$ is the position vector, and t is the time.

Both (1) and (2) are statements in classical physics. Part (2) obviously involves differentiation, or infinitesimal calculus. This concept was invented by Newton and, independently, by Leibniz.

(3) Newton also discovered the generalized binomial formula:

$$(x+y)^r = \sum_{k=0}^{\infty} \binom{r}{k} x^{r-k} y^k \text{ , where } |x| > |y|,$$

r is any (real) number, $k! = k(k-1)\cdots(2)(1)$, $\binom{r}{0} = 1$,

and $\binom{r}{k} = \dfrac{r(r-1)\cdots(r-k+1)}{k!}$. For example, for $1 > |x|$ we have:

$$(1+x)^{-1} = \binom{-1}{0}1^{-1}x^0 + \binom{-1}{1}1^{-2}x^1 + \binom{-1}{2}1^{-3}x^2 + \binom{-1}{3}1^{-4}x^3 + \cdots =$$

$$= 1 + \frac{-1}{1}x + \frac{(-1)(-2)}{2!}x^2 + \frac{(-1)(-2)(-3)}{3!}x^3 + \cdots = 1 - x + x^2 - x^3 + \cdots.$$

No doubt, the most profound mathematical contribution by Newton is the discovery of differentiation. He called it fluxion and defined it by means of a complicated notation that is not in use anymore. The modern notation that we see in (2) is due to Leibniz (and it is called the Leibniz notation of the derivative).[1]

---

[1] In the context of the thesis that the ultimate sources of groundbreaking mathematical ideas are transcendental (Section 8.4), the claim that one of Newton and Leibniz has priority over the other in the invention of calculus is a moot point.

10.2 A SAMPLE OF LEIBNIZ'S MATHEMATICS

(1) As we have mentioned earlier, Leibniz co-invented with Newton infinitesimal calculus. His (and the modern) notation for the derivative of the function y with respect to x is $\frac{dy}{dx}$. Notation is very important in mathematics, for a good notation makes it much easier to efficiently express theories.

(2) $\begin{vmatrix} a & b \\ c & d \end{vmatrix} = ad - bc$. *Determinant* is a basic notion in mathematics. Leibniz defined and used general determinants: if the entries of an $n \times n$ matrix A are $a_{i,j}$, $i = 1,2,...,n$, $j = 1,2,...,n$, then $\det(A) = \sum_{\sigma \in S_n} \left(\text{sgn}(\sigma) \prod_{i=1}^{n} a_{\sigma(i),i} \right)$, where S_n is the group of permutations of $1,2,...,n$, σ ranges through the elements of S_n, and $\text{sgn}(\sigma)$ is 1 or –1 depending on whether σ is an even or an odd permutation, respectively.

There are many other results and ideas that can be extracted from Leibniz's (and Newton's) huge collection of notes, essays, and letters, and some of them were way ahead of their time. Consider, for example, the following enigmatic theorem:[2]

(*) *A contains B, and B contains C, therefore A contains C.*

He calls it an axiom, yet he proves it. From his proof, it transpires that 'A contains B' means $A = AB$. Translating (*) into modern language and notation, it becomes a simple exercise in elementary semigroup theory, where a semigroup is an algebraic structure with an associative operation: for every a,b,c in a semigroup, we have $(ab)c = a(bc)$. Here is (*) again, with a mini-proof.

(**) If A,B,C belong to a semigroup, if $A = AB$ and if $B = BC$, then $A = AC$.

---

2 From his essay 'Principles of a Logical Calculus', 1679 (found in Wiener's selection).

Proof: $A = AB = A(BC) = (AB)C = AC$.

We see here some intrinsic difficulties in doing mathematical research during Leibniz's time: he and Newton did not have the efficient language and notation of modern mathematics.

10.3 ISAAC NEWTON: A SHORT BIOGRAPHY

Newton was born in Lincolnshire, East Midlands of England, on December 25, 1642 (Old Style dates). His father, who was a yeoman farmer, died before Newton was born. When Newton was three, his mother remarried and moved away from the farm, leaving him with his grandmother. Little Isaac never experienced genuine parental love; his grandmother did not show much affection to him either. Likely because of his unhappy childhood, Newton became a recluse and remained such throughout his entire life.

When Isaac was about seven, his stepfather died, and his mother together with her three small children from her second marriage, moved back into the farm. At 12, Newton was sent to live with another family, closer to his school.

When he was 16, his mother decided that he should come back and start running the farm. A few months later, it was clear that Newton's destiny was not to be a yeoman: he neglected the farm and spent most of his time reading books and constructing various ingenious mechanisms. His mother prudently decided that it would be most sensible for him to continue his education. In 1661, he was admitted to Trinity College, Cambridge, where he got his Master of Arts degree four years later. That same year the great (bubonic) plague ravaged England; the university closed, and Newton returned home, where he worked on fluxion (calculus), optics, and laws of gravitation. In an extant manuscript from November 1665, he used differentiation to find the tangents and the radii of curvature at all points of a curve. He published his ideas many years later. His nine-page manuscript 'On the Motion of Bodies in Orbit', written in Latin, where he uses *fluxion* to study the orbits of the planets, was published in 1684.

Newton never cared much about publishing his ideas and results. He considered that to be an act of vanity and a call for socializing, both of which he resented. He wrote in a letter to the mathematician John Collins, where he states empathically,[3] 'I see not what there is desirable in publick

---

[3] [Los], p. 103.

esteeme.' Once he was asked to edit the notes of a translation of a certain algebra book; he consented but only on condition that his name should not appear at all in the publication.[4]

In 1666, he returned to Cambridge where he was elected a Fellow of Trinity and continued his spectacularly fast ascent through the academic hierarchy of his time to reach the peak only six years later when he became a Fellow of the Royal Society.

Newton was also not very eager to bequest his knowledge to his students. He avoided giving lectures to students, even though Fellows was required to give talks every week. Neither was he a charismatic lecturer; his talks were attended by only a few students, and sometimes nobody showed up.

His *magnum opus* was the book *Philosophiae Naturalis Principia Mathematica* (*Mathematical Principles of Natural Philosophy*). Written in Latin, it was published in 1687. The laws described in this book would go down in history as Newton's laws of motion.

Newton never married. In fact, he never had any meaningful relationship with a woman. When his friend John Locke tried to intervene, Newton rebuffed him angrily. He wrote a letter where he accused Locke that he 'endeavoured to embroil me with woemen.'

His attitude towards women may have also been shaped by his moral scruples. He kept a list of sins that he had succumbed to, and 'lust' never showed up in that list. Instead, we find innocuous items there, such as 'eating an apple in the church' and 'neglecting to pray'.

After Newton died, it was discovered that he had been deeply involved with alchemy. His personal library contained 169 books on alchemy, and his notes on that subject contained around one million words. It also transpired that he devoted much time to alchemical experiments: in the short span of five years between 1679 and 1684 he wrote 15,000 words of comments on these experiments only. Alchemy was Newton's pathway to mysticism. More about that in a couple of sections.

Newton died in Kensington, on March 31, 1727 (Old Style), aged 84.

10.4 WILHELM LEIBNIZ: A SHORT BIOGRAPHY

Leibniz, like Newton, never married. He also discovered infinitesimal calculus. His father also died relatively young. And Leibniz was also an alchemist.

---

[4] [Ball], p. 323.

Gottfried Wilhelm Leibniz was born on July 1, 1646, in a relatively affluent family. His father Friedrich Leibniz was a professor of moral philosophy at the University of Leipzig. Friedrich died when Gottfried was six. He left a large personal library, with most of the books in Latin, that Gottfried started using from the age of seven. In a way, and unlike Newton, Gottfried Leibniz's academic career was already traced in his early childhood.

When he was 14, Leibniz enrolled at the University of Leipzig; his bachelor's degree was attained only a year later, and he received his master's degree in 1664 when he was 18. The same year he successfully defended a dissertation in philosophy.

In 1666, Leibniz started his study of law at the University of Altdorf. The same year (!) he defended his doctorate in law. After that he declined an academic position at Altdorf; instead, he became an alchemist! More precisely, he became a secretary of the Nuremberg Alchemical Society. According to some sources,[5] the 'alchemical society' was in fact the Brotherhood of the Rosy Cross (*Fraternitas Roseae Crucis*), the secret society of initiated sages and their disciples. Alchemy (but not the 'profane' alchemy) was one of the mystic pursuits of this society. We will say more about alchemy later on in this chapter, and we will briefly cover the society of Rosicrucians in Section 12.2, in relation to René Descartes.

Leibniz's involvement with alchemy seems to have been far less considerable than Newton's. The lack of information may be due to the strict secrecy of initiated alchemists. What we do know is that his variable interest in this subject lasted his entire adult life: according to some reports, even on his deathbed, he inquired about the progress of some alchemical experiments he had been financing. We also claim that his familiarity with alchemy influenced his philosophy; more about that in a couple of sections.

In 1667, Leibniz got the position as a legal advisor to the Elector of Mainz and Archbishop of Frankfurt, which he held until approximately 1674. In 1673, Duke Johann Friedrich (John Frederick) of Brunswick, with whom Leibniz corresponded extensively, offered to Leibniz the post of counselor. Leibniz, who was at the time in Paris, postponed accepting the offer until 1675. Duke Johann Friedrich was the first of the three consecutive rulers of the House of Brunswick in Hanover who sponsored Leibniz. During the following 41 years until his death, Leibniz was their legal

---

[5] Papus: *Tarot of the Bohemians*, 1919 (Found in [Hall], p. 450.).

counselor, historian, political adviser, diplomatic emissary, and librarian of the extensive ducal library.

Unlike Newton, Leibniz didn't produce any single *magnum opus*; his writings were dispersed throughout many essays, notes, and letters. The volume of his correspondences is formidable: he had over 1,000 correspondents from many places (including China!) with whom he exchanged around 15,000 letters throughout his lifetime. On top of that, he produced a prodigious number of essays on various topics, including mathematics and philosophy.

Leibniz died in Hanover on November 14, 1716, aged 70.

10.5 NEWTON'S AND LEIBNIZ'S MYSTICISM

There was Newton-the-scientist and Newton-the-alchemist – two almost disjoint personalities. As far as we know, it seems that Newton's mysticism was almost entirely confined to alchemy-related matters. We will devote the next section to it. On the other hand, Leibniz's mystical interests and activities were spread throughout a variety of disciplines, including his philosophy (Section 10.7). So, this section will mostly be about Leibniz.

There are but a few hints of mysticism in Newton's scientific opus. Newton wrote in *Principia Mathematica* that 'matter [is created] in [] solid, hard, impenetrable, and moving particles'. Leibniz disagreed. However, as we will see in the next section, Newton-the-alchemist, who was not known to Leibniz, also disagreed with Newton-the-scientist.

For Newton-the-scientist, rigid time and space provided immutable coordinates for the solid material reality. Leibniz took issue with Newton's mechanistic concepts of gravitation, matter, time, and space. According to Leibniz:[6] 'Strictly speaking, space, time, causation, material objects, among other things, are all illusions.' We earlier have encountered quite a few mathematicians (Poincaré, Cantor, Brouwer) who concurred.

So do true mystics: for them, there is a simultaneity of all events in time and space. The great mystic and poet William Blake wrote the poem 'Milton' in response to Newton's concepts of time and space; in it, he stated that, 'Satanic Space is delusion.' According to Blake, Newtonian science is superstitious nonsense, and the only reality is imagination.[7]

---

[6] https://iep.utm.edu/leib-met/.
[7] [Gui].

Newton, 1795-c. 1805 (William Blake). Blake was also a talented engraver and painter: this is his depiction of Newton

As we travel backwards in time starting from the present, the concept of 'miracle' becomes wider in scope, as well as more acceptable. No wonder then that both Newton and Leibniz acknowledged their existence. Newton:[8] 'For Miracles are so called not because they're the work of God but because they happen seldom and for that reason create wonder.' Leibniz:[9] 'Miracles are quite within the order of natural operations.'

Leibniz didn't merely postulate the existence of miracles: he actually collected cases of miracles and mystical experiences.[10] Moreover, he investigated such cases: In 1691, Duchess Sophia, Leibniz's patron at the time (and, as we will see, whose sister was an important person in Descartes' life), asked Leibniz for his opinion on a certain Rosamund von der Asseburg, who claimed that Jesus Christ appeared to her and that he somehow wrote answers to her questions in previously sealed envelopes. Leibniz met Rosamund personally and evaluated her. His verdict: Rosamund had

---

[8] [Dob], p. 230.
[9] [Wie], p. 298.
[10] [Cou]; article by Daniel J. Cook (*Leibniz and Enthusiasm*).

natural talents that allowed her 'to turn her [religious] ardor into a vision; one should not repress or rebuke her – but preserve her.'[11] It is clear that he took Rosamund's talents very seriously.

Many a historian's claim notwithstanding, there is no doubt whatsoever that Leibniz was much more than a rational mystic; in fact, he may have been a true mystic. Here is a relevant quote from Leibniz's *On The True Theologia Mystica* (ca. 1690):[12] 'there are many who are learned without being illuminated []. The light does not come from without.' It is inconceivable that he would make this claim without himself perceiving 'the light from within'.

There is more: 'When anyone sees the true light, he is convinced that it is of God.' This could only have been written by someone who has seen or perceived the 'true light'.

Further: 'Within our self-being there lies an infinity, a footprint or reflection of the omniscience and omnipresence of God'. And: 'The denial of self is [] the love of the origin of our self-being, or God.' With this Leibniz expressed the most fundamental experiential true mystic edicts!

Leibniz's route into the mystic realms was by meditation (that he called 'quietude'), wherein he induced OBEs (out-of-body experiences; Section 6.5) through thinking about his thoughts! Quoting him:[13]

> It seems to me that when I think of myself thinking and already know, between the thoughts themselves, what I think of my thoughts, and a little marvel at this triplication of reflection, that I turn upon myself wondering and do not know how to admire this admiration.... Anyone who desires a experience of these matters should begin to think of himself and his thinking sometime in the middle of the night, perhaps when he [!] cannot sleep, and think of the perception of perceptions and marvel of this condition of his, so that he comes gradually to turn more and more within himself or to rise about himself, as if by a succession [sequence] of spurts [bursts] of mind. He will wonder that he has never before experienced this state of mind.

---

[11] Ibid, p. 121.
[12] [Lei], p. 367–369.
[13] [Cou]; article by Daniel J. Cook (*Leibniz and Enthusiasm*).

So, he must have spoken from his own experience, when he stated the following: 'I believe that he who will meditate upon the nature of substance, will find that the whole nature of bodies is not exhausted in their [size, figure, and motion], but that we must recognize something which corresponds to soul.' The soul, according to Leibniz, is imperishable; in his short paper 'Confessio naturae contra atheistas', he endeavored to demonstrate the necessity of the immortality of the soul. Moreover, Leibniz maintained that individuality is 'uniformly conserved',[14] thus departing from mainstream Christian dictums.

'There is [] nothing [] dead in the universe', wrote[15] Leibniz. And this brings us to the next two sections.

10.6 NEWTON AND ALCHEMY

Alchemy is not a bundle of recipes on how to make gold by slowly boiling quicksilver (mercury) for 40 days, spiced with sulfur and salt. The claim that the two of the most ingenious people in our historic times fell in for such a banal swindle is ludicrous. Even less believable is the claim that, after trying and failing to get gold by cooking various elements, Newton would be motivated to further deepen his knowledge on the subject by reading 169 books on alchemy and then writing comments that were some one million words long.

So then, what is alchemy?

The very influential psychiatrist, Carl Jung, who was mentioned earlier (Section 8.3), considered alchemy to have been more the predecessor of modern psychology than of modern chemistry! Jung himself was probably not an alchemist; however, he studied the subject in depth and wrote two books about it.

Newton devoted many years, indeed decades, to alchemy. Clearly, *he* thought alchemy was important. During all that long period of time, as we have mentioned, he was scrupulously secretive about his alchemical overtures. His notes were mostly cryptic and hard to decipher. He was seldom explicit and clear, as he was in the following short typically alchemical declaration:[16] 'all things are generable.'

Newton-the-scientist did not allow Newton-the-alchemist into his life. However, sometimes the latter managed to intrude; quoting Newton

---

[14] [Wie], p. 116.
[15] 64[Lei2], p. 266.
[16] [Dob], p. 56.

from his book *Opticks*:[17] 'The changing the bodies into light, and Light into Bodies, is very conformable to the Course of Nature, which seems delighted with Transmutations.'

It is amazing that the first edition of *Principia* (1687) carries the following statement:[18] 'Anybody can be transformed into another, of whatever kind, and all the intermediate degrees of qualities can be used in it.' This was clearly an intrusion by Newton-the-alchemist, or the 'crazy' Newton, and it was promptly purged from the subsequent editions! The sanitizers of scientific history started their cleansing work early!!

In his cryptic and coded alchemical writings Newton was careful to distinguish between 'vulgar chemistry' and 'vegetative chemistry', the former being mechanical, the latter requiring 'vegetable' action[19] and, according to Newton, constituting true alchemy. A vegetable action involves *Azoth* – the mysterious spirit of life – as well as *fermentation* and *illumination*, technical terms in alchemy. Illumination is a 'divine' activating agent in alchemical work. It is closely interlinked with the alchemist's consciousness. This brings us to the main aspect of alchemy.

According to true alchemists, the real aim of alchemy is not the transmutation of metals; the main goal is the transmutation into a higher state of consciousness of the alchemist himself. Alchemy is, above all, a mystic initiation. The fully initiated find within themselves the elusive philosopher's stone and the elixir of life.

The Canadian writer and philosopher Manly P. Hall stated it well:[20] 'The greater alchemy is first taken place within the soul of man.' Chemistry is merely the material body of one of the stages of the alchemist initialization. Mechanical chemistry is woefully inadequate to achieve the transmutation of metals. The transmutation of metals is just a test that experimentally verifies that the perspective alchemist has successfully decoded the enigmatic alchemical literature, permeated with obscure, coded language and allegories. It is a kind of intermediate application of true alchemy.

Alchemical teaching has a philosophical foundation. Indeed, during the Middle Ages, alchemy was philosophy, science, and religion, all in one. The gold-making pursuit was mainly a screen that shielded the alchemists from persecution by exposing them to ridicule and scoffing. To some extent, this continued in modern times too.

---

[17] [Los], p. 125.
[18] [Dob], p. 23..
[19] Ibid, p. 30; original source is Newton's 'Of Nature Laws'.
[20] [Hall], p. 509.

Alchemy is ancient. The first alchemist noted in our linear history or mythology seems to have been the Hindu god Shiva, incarnated or not. Shiva delivered to humanity Tantras, a system of spiritual knowledge. It has been stated in Tantras that Shiva practiced chemistry, which has 'proceeded *from him* [my emphasis].'[21]

Historic European, Near Eastern, and North African alchemy starts with Hermes Trismegistus ('Hermes the Thrice-Greatest'), an ancient Egyptian alchemist. According to some, Khem was an ancient name of Egypt, and both the word alchemy and chemistry are derived from it. The subsequent European gallery of alchemists includes some very colorful personalities, including were Paracelsus (1493–1541), a Swiss alchemist and a miracle-healer, and Count Saint-Germain (~1690–?), the never-aging alchemist, who, over the span of many decades, always looked as if he was in his early 40s. The 18th-century philosopher Voltaire said that Saint-Germain 'never dies and knows everything.'

One of the last major known alchemists was the elusive and ageless Fulcanelli, who was born a few decades before 1916 (some claim in 1877), disappeared in 1926, appeared again a few years later, and of whom it is not known when he died. In 1937, he was interviewed by the French writer Jacques Bergier. Here is an intriguing part of the transcript:[22]

> You're on the brink of success, as indeed are several other of our scientists today. Please, allow me, be very, very careful. I warn you… The liberation of *nuclear power* is easier than you think and the *radioactivity* artificially produced can poison the atmosphere of our planet in a very short time, a few years. Moreover, atomic explosives can be produced from a few grains of metal powerful enough to destroy whole cities. I'm telling you this for a fact: the *alchemists have known it for a very long time.*

This is puzzling: the interview happened in 1937, many years before the invention of the atomic bomb.

Following the interview, Fulcanelli disappeared again; according to one of his disciples, the last encounter with him was in 1953; Fulcanelli looked

---

[21] [Bos], p. 28.

[22] Powell, Neil Alchemy, the Ancient Science, p. 53, Aldus Books Ltd, London, 1976; quoted in https://nextexx.files.wordpress.com/2018/03/fulcanelli.pdf.

like a man in his 50s, much younger than how he looked the previous time his disciple met him, a couple of decades earlier.

Let us now pay attention to the transmutation of metals: is there any scientific corroboration allowing for the *possibility* of such an outlandish phenomenon? The question is linked to the enigma of the structure of the most elementary units of matter: what are these units and what are their properties? According to the string theory of relatively modern physics (and related upshots like superstring theory), everything is made of oscillating strings of energy. The frequency of the vibrations of the strings determines the properties of the associated matter or energy. Transmutation is then simply a change of vibration. If string theory is correct, is it possible to change the vibration of these elementary particles by means of sheer intent?

We have arrived at the phenomena of psychokinesis (or telekinesis) mentioned by Turing: the influence of mind or consciousness over matter without direct physical interaction. The genuineness of psychokinesis has been proven in strictly controlled laboratory experiments with 'astronomically' huge statistical certainty. There are hundreds, perhaps thousands of published (peer-reviewed) articles on the subject. We will only mention the inaugurating research by J. B. Rhine (influencing the outcome of a dice throw),[23] the experiment of Helmut Schmidt (1975), whose subjects affected the randomly generated sequence of ones and zeros, Dean Radin's experiment where the subjects affected trajectories of photons,[24] and William Tiller's (1997) change of pH (the measure of acidity) of water by meditation.[25]

The subjects of the above experiments were ordinary and more-or-less randomly chosen people. It is altogether another story when the experiments were performed with talented, psychic, or spiritually initiated people, when the results are often extraordinary in many ways. Take, for example, the talented psychic Ingo Swann, whom we mentioned in connection to OBE (Chapter 6). In an experiment with the physicists Harold Puthoff and Russell Targ, Swann increased and decreased, at will, the magnetic field within a superconducting magnetic shield. In another experiment, he 'has been able to deflect the needle of a magnetometer encased in a superconductive substance and buried in concrete.'[26]

---

[23] See, for example, [Rhi2].
[24] See [Rad].
[25] See, for example, [Tag].
[26] [Mos], p. 126.

On top of all of that, and perhaps surprisingly, there exist scientific transmutation experiments with positive results! For four years in the late 1950s, the French scientist Corentin Louis Kervran performed extensive research on various living organisms. In his book[27] *Biological Transmutations*, published in 1966, he gives a detailed account of his work. In the introduction, on page xv, he wrote the following uncompromising statement:[28]

> Early in 1959, after several years of systematic research, I decided to publicize my conviction, indeed my certainty, that there is a hitherto unknown property of matter which has been widely but unknowingly utilized. This new property, which I had demonstrated as a result of thousands of relevant analyses, gives living bodies the ability to transform not only molecules (which is within the field of chemistry), but atoms themselves. In fact, there is a transmutation of matter; a passage of one 'simple body' to another, of one atom to another.

We will leave aside the many outrageous or 'miraculous' outcomes of scientifically controlled experiments in psychokinesis. The above few experiments and the corresponding results, if taken seriously, should be sufficient to postulate the *possibility* that some spiritual adepts, in particular alchemists, have been able to affect the vibrations of the units of matter in the form of hypothetical strings of energy, thus affecting transmutation of matter. The *possibility* that our collective thoughts and expectations configure our perceived world is then just around the corner. And from that vantage point one can spot on the horizon Brouwer's gargantuan claim that the universe is 'freely built by us'.

We mention that various experiencers, including mystics, have reported for centuries that there is a change of frequency associated with changes in their awareness of the perceived reality. We use the occasion to digress into the 'mystic string theory' and, at the same time, link Brouwer to Newton through a chain of personalities and events. In the year 1870, the American physician and scientist Edwin D. Babbitt (1828–1905), one of the discoverers of chromotherapy (color therapy), 'commenced cultivating, in a dark room and with closed eyes, [his] interior vision.' After a

---

[27] [Ker].
[28] [Ker], p. xv.

few months, he managed to *see* (clairvoyantly) the structure of the most elementary subatomic particles of matter. Accordingly, they are multiple strings of energy linked in knotted spirals; each of these energy spirals is made of smaller spirals winding around them; each of the smaller spirals is made of even smaller spirals winding around them; and this is repeated a total of seven times.[29] See the two figures below.

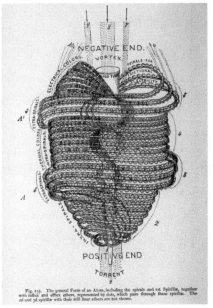

Fig 133. Piece of Atomic Spiral with 1st 2nd and 3rd Spirillæ.

Subatomic strings according to Babbitt

In 1895, the clairvoyant Charles Leadbeater was requested to try to see, also clairvoyantly, the structure of hydrogen. He (allegedly?) succeeded and described it as a certain configuration of linked spirals that were virtually identical to Babbitt's.[30] Should this be dismissed as occultish junk? Perhaps not! The same experiment in clairvoyance produced a description of some atoms that was subsequently verified: Leadbeater *saw* that a hydrogen atom could have one, two, or three 'particles' in its nucleus, and still be hydrogen. Neutrons (the 'particles') and the isotopes of hydrogen (deuterium with two neutrons, and tritium with three neutrons) were discovered some 35 years later (early 1930s).

---

[29] [Bab].
[30] [Bes].

Charles Leadbeater was a prominent theosophist who spent long periods of time in India. The last time he returned to England, he and his
fellow theosophist, Annie Besant, brought to England a young, spiritually
talented Indian boy who was to be groomed to become *Maitreya*, the fifth
and final Buddha. He followed this path until he was 35. Then he experienced a spiritual shift and became a prominent philosopher. His name was
Krishnamurti; we mentioned in Chapter 7 that Brouwer was influenced by
Krishnamurti's philosophy and that he attended his lecture.

10.7 LEIBNIZ, MONADOLOGY, ALCHEMY, AND LEVITATION

Newton:[31] 'All matter [] is attended with signs of life.' Leibniz:[32] 'There is
[] nothing [] dead in the universe.' The obvious congruity of their ontological views is not a coincidence: no doubt it is at least partially due to
their common interest in alchemy. One of the axioms of alchemy is that
'Within everything is the seed of everything.' This seed is not merely living; it is God Himself! In its fractal-like structure, the divine consciousness is reflected over and over again into smaller and smaller parts. Here
is, again, Manly Hall:[33]

> In each grain of sand, each drop of water, each tiny particle of
> cosmic dust, are concealed all the parts and elements of cosmos in
> the form of tiny seed germs so minute that even the most power-
> ful microscope cannot detect them. [] These seeds await the time
> assigned them for growth and expression.

Leibniz went further than merely postulating the omnipresence of life: he
conceived a philosophical theory in which he conjectured the properties
of the most elementary units of substance. He called it *monadology*, bor-
rowing the word *monad* from Pythagoreans (monad is a Greek word for
unity). In Leibniz's scheme, a monad is the smallest, *living*, unit of matter.
'The Monad [] is nothing else than a simple substance, which goes to make
up composites; by simple, we mean without parts. [] [The] Monads are
true atoms of nature, and, in fact, the Elements of things.'[34] Matter in itself
has no substance according to Leibniz:[35]

---

[31] [Dob], p. 19.
[32] [Lei2], p. 266.
[33] [Hall], p. 501.
[34] [Lei2], p. 251.
[35] [Cou], p. 55.

> Indeed I do not take away body, but I reduce it to what it is, for I show that corporeal mass [] is not a substance but a phenomenon resulting from simple substances [monads] which alone have unity and absolute reality.

Monads have a kind of perception; quoting Leibniz from his *monadology*: 'If we wish to designate by soul everything which has perception [], all simple substances or created monads could be called souls.'

They also manifest some sort of emotion; quoting Leibniz from his letter to de Montmort, 1715:[36] 'Monads are subject to passions.'

The monads, being equipped with a kind of consciousness, can interact with other consciousnesses or minds. Powerful or divine minds can supply motion to them, and by extension, motion to matter in general. Leibniz is very explicit about that:[37] 'For [divine] mind supplies motion to matter []. Matter is devoid of motion per se. Mind is the principle of all motion, as Aristotle rightly saw.' This, of course, is in complete defiance of Newton's gravitational theory. We stay with this thesis for a while.

When theories, philosophical or mathematical, are being made, their creators almost always have some specific models in mind. Leibniz, being both a philosopher and a mathematician, is likely not an exception to this rule. His monadology is a theory that has material reality as its originating model. Consequently, for this theory and its model to be compatible, it is necessary to accept that material reality also allows and includes gravitation-defying phenomena! What kind of such phenomena might Leibniz have had in his mind?

Enter Duke Johann Friedrich, Leibniz's first patron. The two had a close relationship until the duke died in 1679. Duke Friedrich converted from Lutheranism to Catholicism in 1651; he was the only Catholic in his family. The reason for the conversion was very unusual. We will now enter a no-go zone for the orthodox scientificology.

In 1649, Duke Friedrich visited Assisi, in Italy, and attended Catholic Mass. During the Mass, the Franciscan friar Giuseppe, or Joseph, of Cupertino, levitated! He *flew* backwards, and he did it twice in duke's presence! The duke was so impressed that he came again the following year and officially made a 'confession of fate', that is, he converted to Catholicism. The rite was performed by Joseph.

---

[36] [141Wie], p. 189.
[37] [LodP], p. 36.

Before we pay attention to Joseph and the phenomenon of levitation, we return to Leibniz to make a point regarding the locomotive responsiveness of his monads to influences of (divine) mind. Friedrich's experience was known to him. He confirms duke's account[38] by laconically stating that the duke was converted to Catholicism by the miracle-working Father Joseph. Further, Leibniz admired the Spanish nun, mystic, and author Theresa of Avila (1515–1582), who was canonized to sainthood in 1622. Theresa also levitated several times during prayer-induced raptures! It is certain that Leibniz accepted the genuineness of the phenomena, and it is reasonable to postulate that his theory of monads being movable by the divine mind took these phenomena into account.

Joseph of Cupertino may have been the person who levitated the most according to our linear recorded history, with the possible exception of the Indian mystic Milirepa (1052–1135). Most of Joseph's levitation episodes occurred in Assisi, where he was a friar for nine years. It is interesting to note that more than four centuries earlier, St. Francis of Assisi (ca. 1181–1226, canonized 1228) also levitated in the same Basilica. There are some 230 Catholic saints to whom this feat is attributed.

Unlike Milirepa's long-distance flights, Joseph's levitations were involuntary; indeed, he was often embarrassed when he floated in the air in the middle of the Mass. When he levitated, he was stiff as dead; even his clothes, including his tunic, stayed undisturbed as if he was a still holographic image superimposed over the 'normal' reality. Apparently, the mind that moved his body was external to him. The levitations at times lasted for an hour or so, and a few times they happened several times during a single day. At times Joseph flew some 30 meters high. There were many witnesses, including some noblemen, and most of the time the events were perceived simultaneously by large groups of people, thus virtually excluding explaining the events as hallucinations. Joseph died in 1663; when some suggested that he should fight the impending demise of his body and pray for his own health he responded laconically, 'God forbid!'.

The expected reaction of skeptics to the phenomenon of levitation would be to pose the rhetorical question: how could people take such an obvious impossibility seriously? Short answer: because the genuineness of the levitation phenomenon has been virtually confirmed by numerous scientific experiments. And because in the context of the paranormal

---

[38] [Gros], p. 84; original source [Wie], p. 9.

phenomena that we have (partially) set up so far, specifically of telekinesis, levitation is not a miracle anymore. On the side, there is a lot of irony and some satisfaction in considering seriously gravitation-defying events and experiments in the chapter devoted to Newton.

There exist thorough investigations of various talented mediums who could levitate, especially during the second half of the 19th century: such were the experiments by the chemist William Crookes with D. D. Hume, and by the physicist and astronomer Johann Zölner with Henry Slade. These and similar experiments are casually dismissed by the followers of scientism as lacking scrupulous controls, or by declaring the experimenters as incompetent. Martin Gardner, covered in Section 6.2, wrote,[39] 'Any magician will tell you that scientists are the easiest persons in the world to fool.'

I trust naïve scientists more than conceded magicians. However, even if we dismiss such experiments as spiritualistic mumbo-jumbo, the assessment of the authenticity of gravitation-defying phenomena produced in scientific laboratories is not affected by much.

Rather reluctantly I must restrain myself from the temptation to expand much further. For the curious, the following books contain (ad hoc) experiments with various cases of levitation: Alfred Russel Wallace: *Miracles and Modern Spiritualism* (1878), Louis Jacolliot: *Occult Science in India and Among the Ancients* (1884), Theodore Illion: *In Secret Tibet* (1937), W. Y. Evans-Wentz: *Tibet's Great Yogi Milarepa, A Biography for the Tibetan being the Jetsün-Kahbum or Bibliographica History of Jetsün-Milirepa* (1951), Brad Steiger: *The Psychic Feats of Olof Jonsson; The Authorized Biography of a Remarkable Sensitive* (1971), Thelma Moss: *The Probability of the Impossible; Scientific Discoveries and Explorations in the Psychic World* (1976), Steve Richards: *Levitation; What It Is – How It Works – How To Do It* (1980), and Michael Grosso: *The Man Who Could Fly* (2016).

Sticking with our protagonists, here is an account by Sophia De Morgan, where she describes an impromptu séance in a friend's house:[40]

After sitting some time we were directed by the rapping to join hands and stand up round the table *without touching it* [original emphasis]. All did so for a quarter of an hour, wondering whether

---

[39] [Gar], p. 92.
[40] [DeM], p. 26.

anything would happen, or whether we were hoaxed by the unseen power. Just as one or two of the party talked of sitting down, the old table, which was large enough for eight or ten persons after the manner of a lodging-house, moved entirely by itself as we surrounded and followed it with our hands joined, went towards the [skeptical] gentleman out of the circle, and literally pushed him up to the back of the sofa, till he called out, 'Hold, enough.'

This whole episode is worthless from the scientific point of view: materialistic scientists abhor anecdotes, especially those that directly contradict their fundamental axioms. Now, compare Sophia's account with the following:

[During] the 11th meeting, [] the table, instead of merely tilting or rocking on two legs, as it has done so far, rose clear from the floor. The explanation of unconscious muscle action was suddenly no longer applicable, since one cannot push the table up in the air, either consciously or unconsciously, when the hands are on top of it.

Gradually the movements [of the table] became bolder and the lamp was lit for longer periods. By its red glow we could clearly see our hands on top of the table. [] Because the levitations were not too high, I said: 'Come on – higher!' at which the table rose up chest high and remained there for eight seconds. [] At one point the table levitated and floated right across the room: we had to leave our seats to follow it; it appeared to be about five inches off the floor, and the signal lamp remained alight until we crashed into some other furniture.

Is this account of yet another group of spiritualistic suckers? Not at all! The above is a brief summary of one out of 200 scientifically controlled experiments, performed over a period of 18 months. The principal investigator was British psychologist Kenneth James Batcheldor (1921–1988), and the results were published in the journal of the Society of Psychic Research ('Report on a Case of Table Levitation and Associated Phenomena', 1966). The sitters were randomly chosen volunteers! A scientifically controlled table-levitation experiment with positive results? Isn't that amazing?

There is one very interesting feature of these experiments that must be noted: out of 200 sittings, 80 were held with a certain Mr. W. G. Chick, and these were the *only* occasions when phenomena occurred. According to Batcheldor, 'positive results were almost guaranteed in the presence of W. G. C., even if there were only one more sitter [Mr. Chick and himself]'. So, it seemed that it was necessary to utilize the talents of at least one accidental medium to get positive results! One consequence would be that the scientific requirement of repeatability of the experiment is not guaranteed; a Mr. Chick is required as a channeler from the ultimate source of these phenomena to our sensual reality. We will call this requirement 'Chick condition'.

Yet, amazingly, about four years after Batcheldor's article was published, the results were replicated under similar conditions! C. Brooks-Smith and D. W. Hunt[41] assembled a group of four people, who knew each other well and who were interested in telekinetic phenomena. They were set up in 'laboratory conditions', under full lighting and with cameras at hand.

> Sitters at the table rested both hands gently on the table top, in full lighting conditions. [...] Normal conversation was carried on during the sitting but no doubt each person, in his [sic! one of them was female!] own way, was willing or wishing for the table to make a movement. Results were in fact rapidly produced: knocks and raps were heard, apparently from the table, at the first meeting. Tilting of the table occurred at the second meeting, and after a few sessions, the phenomena had developed into violent table movements over which no exact control seemed possible, and which indeed caused anxiety due to the possibility of injury. [] These movements were produced at early stages of successive sessions and included the rising of the table some five or six feet clear of the floor, its movement over the whole of the room whilst in the air and a peculiar oscillating descent to the floor. During all of the movements the experimenters as far as possible maintained a light one-finger contact with the table, but these were unavoidably lost on occasion and many movements were possibly made without contact.

---

[41] [Bro]*.

Further, from Sitting No. 3:

> Almost at once the table became levitated and remained in a gently floating state. On one occasion when all hands were withdrawn simultaneously, the table jumped still further rather violently and hands were promptly replaced. A repetition of this procedure shortly afterwards produced moderately [severe] jumps quickly followed by the descent of the table to the floor.

It's hardly worth mentioning that these two fascinating articles were almost completely ignored both by the scientific establishment and the public at large.

Assuming we take these experiments seriously, what to make out of them? We recall our earlier postulate where we stated that consciousness could exist without material basis, and where we have stipulated the existence of native realms, or planes, of such incorporeal consciousnesses. In this context, the following three explanations of the levitation phenomenon seem the most straightforward:

- Mechanical explanation: with some help from the channeler, the incorporeal consciousnesses manifest hitherto unknown anti-gravitational forces

- Mystical explanation: incorporeal consciousnesses manipulate the (unknown) parameters that determine the structure of the table

- Psychological explanation: incorporeal consciousnesses affect our 'objective' reality by manipulating our consciousness

The truth may be some combination of the above three.

CHAPTER 11

The Multitalented Blaise Pascal

Truth is so obscure nowadays and lies so well established that unless we love the truth we shall never recognize it.

PASCAL

11.1 THE MANY TALENTS OF BLAISE PASCAL

Blaise Pascal (1623–1662) was a French mathematician, scientist, inventor, philosopher, writer, and accidental mystic, all in one. Hence, his contributions are spread over many scientific fields: there is a unit of pressure named pascal in his honor; it is defined to be one newton per square meter. There is a computer programming language also called Pascal: when he was almost 19, Pascal constructed a calculator, later named Pascaline, for addition and subtraction.

DOI: 10.1201/9781003282198-11

Blaise Pascal

His most significant contribution to mathematics was his development of probability theory. His name is also remembered through the Pascal triangle: a scheme for recursively computing the coefficients in binomial expansions.

$$\begin{array}{c} 1 \\ 1\ \ 1 \\ 1\ \ 2\ \ 1 \\ 1\ \ 3\ \ 3\ \ 1 \\ 1\ \ 4\ \ 6\ \ 4\ \ 1 \\ 1\ \ 5\ \ 10\ \ 10\ \ 5\ \ 1 \\ \vdots \end{array}$$

$$(x+y)^0 = 1$$
$$(x+y)^1 = x+y$$
$$(x+y)^2 = x^2 + 2xy + y^2$$
$$(x+y)^3 = x^3 + 3x^2y + 3xy^2 + y^3$$
$$(x+y)^4 = x^4 + 4x^3y + 6x^2y^2 + 4xy^3 + y^4$$
$$(x+y)^5 = x^5 + 5x^4y + 10x^3y^2 + 10x^2y^3 + 5xy^4 + y^5$$
$$\vdots$$

Left – the Pascal triangle: each number in a row is the sum of the adjacent numbers in the preceding (higher) row. Right – binomial expansions: the coefficients (called, not surprisingly, binomial coefficients) are equal to the corresponding numbers in the Pascal triangle.

In general,

$$(x+y)^n = \binom{n}{0}x^n + \binom{n}{1}x^{n-1}y + \binom{n}{2}x^{n-2}y^2 + \cdots + \binom{n}{k}x^{n-k}y^k + \cdots + \binom{n}{n-1}xy^{n-1} + \binom{n}{n}y^n,$$

where each binomial coefficient is $\binom{n}{k} = \dfrac{n!}{k!(n-k)!}$ (the factorial ! is defined by $n! = n \cdot (n-1) \cdots 3 \cdot 2 \cdot 1$). The relationship of the numbers in the Pascal triangle can then be described as follows: $\binom{n}{k} = \binom{n-1}{k-1} + \binom{n-1}{k}$.

The number of combinations of k elements out of n elements is also $\binom{n}{k}$; with this meaning, the binomial coefficients play a significant role in basic probability theory introduced by Pascal.

11.2 PASCAL: A BRIEF BIOGRAPHY

Blaise Pascal was born in 1623 in a relatively well-to-do family. His mother died when he was three, and he and his two sisters were brought up by his father Étienne. Perhaps due to a lack of motherly care and love, Blaise was a sickly child. Eventually, the older sister Gilberte (1620–1687) helped by taking care of Blaise and the younger sister Jacqueline. Gilberte wrote a biography of Pascal. Jacqueline (1625–1661), Blaise's favorite and also a child protégé, eventually became a nun.

Blaise's education was mostly taken care of by his father and by some of his father's acquaintances; Blaise never attended a public school.

Initially, Étienne tried to prevent Blaise from indulging in the 'delights of mathematics', but he relented when the young Blaise showed him his proof of the claim that the sum of the interior angles in a triangle is 180°.

At 16, Pascal wrote a brief essay on conic sections (circle, ellipse, parabola, hyperbola), which he later expanded into a monograph, which is now lost. Surviving is the following intriguing 'mystic hexagram' theorem: if a hexagon is inscribed in a conic section, then the intersection points (if any) of the opposite sides of the hexagon are on a single straight line.

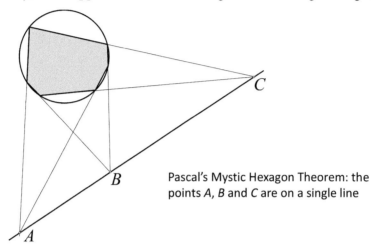

Pascal's Mystic Hexagon Theorem: the points *A*, *B* and *C* are on a single line

When he was 24, Pascal suffered a major health crisis that may have been caused by intestinal tuberculosis.

In 1650, when Pascal was 27, he met Charlotte Gouffier, the 18-year-old sister of a friend. This proved to be fateful for both: it seems that Pascal fell in love with Charlotte and that she was not indifferent either. However, their affection did not evolve conventionally, and Pascal's feelings seem to have been hurt. He came to hate every affection and went as far as to censure his sister Gilberte for caressing her small children.

Pascal and Charlotte exchanged surreptitious letters for many years. Pascal tried to convince her to join a nunnery; he wrote gravely, 'The church has been weeping for you for sixteen hundred years.'[1] Charlotte eventually joined a nunnery for a short period of time; her mother managed to extract her and forced her back into a mundane life. A year after Pascal died, and at the ripe age of 32, she married. Her life was rather tragic: two of her children died young, and the third one was, in the language of her times, a cripple.

Following a mystic episode that happened in 1654 – to be covered in the following sections – Pascal made a sharp turn in his life by joining Port Royale Abbey, a stronghold of solitary men with Jansenist affinities.

---

[1] [Bis], p. 50.

Jansenism was a theological movement within Catholicism that put stress on original sin (humans are born with sinful inclinations) and, as a consequence, human depravity. Jansenism fit well with Pascal's dominating emotional and mental state. While Pascal never officially converted to Jansenism, he was de facto their spokesman during the many theological disputes with the Jesuits. However, under his influence, both of his sisters did convert to Jansenism. The teaching of Jansenism was outlawed by the pope in 1713.

Pascal did not give up mathematics after 1654: in 1659, he published a few essays on cycloids. Leibniz copied it when he was in Paris, and it was subsequently used by him in the invention of calculus.

In the last few years of his life, Pascal was sick; there exists no definite diagnosis of his illness. What is known is that he was weak and mostly in pain, which he tried to dilute by doing mathematics. According to a friend, Pascal was 'in a state of languor [] and living on ass' milk and bouillions [soups].'[2]

The year he died he instructed his sister Gilberte not to be sorry for him. He said[3] 'Sickness is the natural state of Christians because therein one is [] in suffering, illness, [] the deprivation [] of all pleasures of the senses' He died on August 19, 1662, from a hemorrhage of the brain.

11.3 DECODING THE MEMORIAL

A servant in charge of funeral duties noticed a bulge in Pascal's coat lining. He cut it open and found a frantically scribbled piece of paper and a parchment with the same text carefully transcribed, with a few additions. The note, Pascal's *Memorial*, was about an event that had happened eight years before his death; it was published nine years after Pascal's death.

Here it is:

> Monday, 23 November, feast of St. Clement, pope and martyr, and others in the martyrology. Vigil of St. Chrysogonus, martyr, and others. From about half past ten at night until about half past midnight,
> FIRE.

---

[2] [Bis], p. 87.
[3] Ibid, p. 94.

GOD of Abraham, GOD of Isaac, GOD of Jacob not of the philosophers and of the learned. Certitude. Certitude. Feeling. Joy.
Peace. GOD of Jesus Christ. My God and your God. Your GOD
will be my God. Forgetfulness of the world and of everything,
except GOD. He is only found by the ways taught in the Gospel.
Grandeur of the human soul. Righteous Father, the world has
not known you, but I have known you. Joy, joy, joy, tears of joy.
I have departed from him: They have forsaken me, the fount of
living water. My God, will you leave me? Let me not be separated
from him forever. This is eternal life, that they know you, the one
true God, and the one that you sent, Jesus Christ. Jesus Christ.
Jesus Christ. I left him; I fled him, renounced, crucified. Let me
never be separated from him. He is only kept securely by the ways
taught in the Gospel: Renunciation, total and sweet . Complete
submission to Jesus Christ and to my director. Eternally in joy
for a day's exercise on the earth. May I not forget your words.
Amen.]

At the superficial level, Pascal's *encounter* with Jesus Christ is similar to
Ivan's experience (Section 1.1). However, there are major differences: Ivan's
experience is easily classifiable as a hypnagogic hallucination; not so for
Pascal's vision. The primary reason is the length of the episode: hypnagogic
hallucinations usually last a few seconds, and in some exceptional cases, a
few minutes. Pascal's episode was the full two hours. So then, what was it?

There are three obvious possibilities: either it was a near-death experience, it was a spontaneous vision triggered by Pascal's health and his
mental state, or it was induced by meditation or by some other ad-hoc
technique causing sensory deprivation (for example, repetitive chanting).

The third option is unlikely: had Pascal stumbled upon some kind of
technique for achieving altered states of consciousness that enabled him
to perceive a manifestation of *God*, this would not have been a once-in-a-
lifetime experience. Besides, he never mentions deliberate meditations or
anything of that sort in his prolific writings.

In order to scrutinize the remaining two possibilities further, we back
up to September of the same year (1654), when Pascal had a long conversation with his younger sister Jacqueline, who, as we remember, was a nun.[4]

---

[4] The original source is Gilberte's biography of Pascal; [Bis], p. 53.

He told[5] Jacqueline that 'he [felt] a great abandonment on God's part', and that he thought that he relied too much on 'reason and his own mind [] rather than the movement of God's spirit.' Thus, his overall mental state was oriented in the direction of desperation and anguish. As we'll see in the last section of this chapter, this is a fairly typical condition that is conducive to experiencing mystic travels and, in particular, of perceiving visions of Jesus Christ or some other of God's representations.

On the other hand, given the frailness of Pascal's health, we cannot exclude the possibility that Pascal had a near-death experience (NDE): that he virtually died and then came back. The fact that during the vision Pascal experienced 'eternal life' gives some further support to the NDE scenario. We will explore visions and NDEs further in the last section of this chapter.

In both cases, perceiving a being of light (or, as Pascal wrote, FIRE), sometimes identified as Jesus Christ, is a major basic feature of the experience.

It is unfortunate that Pascal did not elaborate further, so we don't know exactly what happened during these two long hours that the vision lasted. It is hugely unlikely that he merely stared for two hours at the image of Jesus Christ; spells of such sort last not more than a few minutes. So, it seems that the encounter involved verbal (or telepathic?) communication, as perceived by Pascal. This hypothesis is supported by Pascal's reaction: 'May I not forget your *words* (original emphasis).'

Pascal was, above all, a devout but rational Christian – rational, because he critically scrutinized and then accepted the main tenants of Christianity. The fact that he carried the written notes of his mystical experience close to his heart for eight years implies that he took the experience very seriously. Hence, he was a mystic too. However, it seems that his mysticism did not go further than the experience described in the *Memorial*.

One of our sources for this chapter is Maurice Bishop's biography of Pascal. In it, Bishop tells us how he was criticized by some historians for sitting on the fence in an earlier essay on Pascal and not giving his opinion of Pascal's mystical experience: was it a genuine event, or a hallucination? Here is his response:[6] 'Very well then. I choose hallucination.'

I don't.[7]

---

[5] [Bis], p. 52.

[6] [Bis], p. 55.

[7] Hallucinations are usually defined as percepts not sourced in the material reality. Since we do not know what the (material) reality is, this definition is essentially meaningless.

11.4 PENSÉES (THOUGHTS)

Pensées is a collection of Pascal's reflections on various topics; it was published after his death. All or almost all entries date from the period after his 'mystic' experience in 1654; the first dateable entry is traced to have been made in September 1656.

Some of his thoughts have a mystic feel:

- *Human life is perpetual illusion…*

- *If a man should study himself first of all, he would see how incapable he is of passing beyond himself.*

Others conflict with some basic mystic postulates:

- *I cannot conceive man without thought; he would be a stone or a brute.*

- *Thought constitutes the greatness of man.*

According to mystic teaching, thinking is an obstacle to achieving true greatness. For example:

- Lao Tzu: *Thoughts weaken the mind.*

- Don Juan:[8] *A man of knowledge doesn't think.*

Pascal strongly supported the assertion that genuine miracles exist. No doubt his vision of Christ went down in his books as one such example. He was also aware of the *miraculous* remission of a sizable and painful swelling at the corner of the left eye of his ten-year-old niece Marguerite. The swelling started when she was seven, and since it also affected the structure of her nose, the girl was scheduled to undergo a medieval-ish surgical removal of the growth by burning it with a red-hot iron. The swelling disappeared instantaneously upon being touched with a reliquary by the mistress of Port Royal Abbey, where Marguerite was a boarder. The miracle was verified by a committee of eight physicians and surgeons; they wrote that the cure[9] 'surpasses the ordinary forces of nature'.

Pascal's acceptance of miracles is reflected in *Pensées*:

---

[8] [Cas2].
[9] [Bis], p. 67.

- *It is not possible to have a reasonable belief against miracles.*

- *Thus those who refuse to believe in the miracles in the present day on account of a supposed contradiction, which is unreal, are not excused.*

- *How I hate those who make men doubt of miracles!*

Pascal's deep resentment of affection borders on misanthropy:

- *All men naturally hate one another.*

- *Man's greatness lies in his knowing himself to be wretched.*

- *No other religion [other than Christianity] has proposed that we should hate ourselves.* This is good according to Pascal, for those who hate themselves 'seek the Being that can really love', which is God.

We end our selections from *Pensées* with a few of Pascal's aphorisms:

- *The eternal silence of these infinite spaces fills me with dread.*

- *God could make it all obvious, but at a terrible price.*

- *The immortality of the soul is a matter which is of so great consequence to us and which touches us so profoundly that we must have lost all feeling to be indifferent as to knowing what it is.*

11.5 NDE AND WAKING VISIONS

Most categorically, true near-death experiences are not concoctions of deluded brains, even though they always happen when the functions of the brain are severely impeded or non-existent. There is circumstantial and scientific evidence supporting this claim.

A strong example of the former is the uniformity of the experience that cuts across the age, religion, and social background of the experiencer, as well as the historical time of the event. A typical NDE starts with an OBE and a rapid glide along a tunnel in the direction of light, followed by an encounter and communication with (light) beings, a panoramic life review, feelings of timelessness, ecstatic joy, and love, and ending with a reluctant return into what we call normal consciousness. These core elements can be found in Plato's account[10] of the NDE of Er, a Greek soldier

---

[10] In Book X of *The Republic*.

who was killed in a battle and subsequently revived, in the Tibetan Book of the Dead, and in modern accounts from across this planet. Why would a soldier of antiquity and a four-year-old North American boy of modern times tell virtually the same story about their NDEs? And why is it blasphemous today, from the point of view of the scientific establishment, to postulate the obvious: that there must be some factual underpinning of the event?

An example of a scientific refutation of the claim that NDEs are products of brains muddled by sickness or drugs is the following study by pediatrician Dr. Melvin Morse. Morse virtually proved that in order to have an NDE, one has to be very close to dying. He interviewed 121 children in a control group, all of whom had very serious diseases (for example, one girl had been completely paralyzed for three months), and a smaller study group of 12 children who had survived cardiac arrest and 'looked death in the face'. Of the 121 children in the control group, *none* had anything resembling an NDE (only three of them had dreams). On top of that, Morse went outside the controlled group and further interviewed 37 children who had been given anesthetic agents, narcotics, Valium, Thorazine, Haldol, Dilantin, antidepressants, mood elevators, and painkillers, to see if drugs could induce an NDE. And again, *none* of them had anything resembling an NDE. So, he asks the rhetorical question, 'If near-death experiences are hallucinations, why did [these patients] not have any experience remotely resembling NDE?' On the other hand, at least eight of the twelve children in the study group had at least one of the NDE traits. We note that the children in the study group were chosen by Morse solely on the basis of their having had a cardiac arrest.[11]

Morse eventually lost the support he had from the 'Hospital's Human Subject Review Committee' and could not continue with his study, despite the positive reviews he got for the (three) articles he published on the subjects of children's NDEs. As he put it, 'the medical establishment wears mental blinders when it comes to the subject of death.'

We now make an exception to our choice not to give anecdotal evidence, and we will briefly describe Morse's very first NDE case: seven-year-old Katie.[12] We do this chiefly because of some similarities with Pascal's *Memorial*. Katie was found floating head down in a pool; she was pulled out but had no 'gag' reflexes when she was attached to an artificial

---

[11] [Mor], p. 21; the original article is [Mor]*.
[12] [Mor].

lung machine for three days. Morse gave her a 10% chance of survival. However, after three days she woke up from the coma and almost immediately recovered. When Katie came for the follow-up check she immediately recognized Morse and gave details about her stay in the hospital while she had been comatose with closed eyes. Morse then asked her an open-ended question: 'What do you remember about being in the swimming pool?' She replied, 'Do you mean when I visited the Heavenly Father?' Morse prompted her to tell him more about 'Heavenly Father', to which Katie replied that she had met Jesus and the Heavenly Father. What she told Morse during their next meeting, a week after, changed his life (his words). Here is the sequence of events. First, she recalled darkness. Then a tunnel opened. Through the tunnel came 'Elisabeth', tall and nice, with golden hair. Elisabeth accompanied Katie to the end of the tunnel, where she met her grandfather and other people, among them two boys, 'souls waiting to be reborn'. The two boys played with her and introduced her to other people. Then Katie was given a glimpse of her home, where she saw her brothers and sisters playing with toys and her mother preparing dinner. After that, Elisabeth took Katie to meet the Heavenly Father and Jesus. The Heavenly Father asked her if she wanted to stay and Katie started crying and said yes. Jesus (according to Katie) then asked her if she wanted to see her mother. When Katie answered in the affirmative, she woke up from her coma.

It is possible that Pascal had an NDE; however, it is more likely that he had a waking vision.

Neither NDEs nor waking visions, especially the latter, are easily testable under controlled laboratory conditions. What we have is mostly the evidence of authenticated or veridical anecdotes. As we have mentioned many times, we will not go far in that direction. We will only narrate a waking vision that strongly resembles Pascal's experience.

The family of Sadhu Sundar Singh (1889–1929) was Sikh, but his mother was open-minded. She had read old Vedic texts to him since his early childhood (he knew Bhagavad Gita by heart when he was seven), and she took him to various sadhus (holy men) and priests in search of the right way for him. He practiced yoga, and he read holy books of many religions, but not the Bible. He considered Christianity to be a 'false' religion. So, he burned the Bible in the presence of his father and participated in the stoning of Christian preachers. But he could not 'get any [spiritual] satisfaction and peace', which brought him to the verge of committing suicide. Then something happened; we quote him:

> Three days after I had burnt the Bible, I woke up about three
> o'clock in the morning, had my usual bath, and prayed, 'O God,
> if there is a God, will thou show me the right way or I will kill
> myself.

He wanted to place his head on a railway track in front of the 5 o'clock
train. At 4:30, he saw 'great lights', making him think that something was
on fire. Then he saw a man, whom he described without any hesitation as
Lord Jesus Christ. 'I heard a voice saying in Hindustani, "How long will
you persecute me? I have come to save you; you were praying to know the
right way. Why do you not take it?"'. He fell and felt 'wonderful peace'.
Sadhu converted to Christianity and became a preacher.[13]

There is indeed a striking similarity between Pascal's and Sidhu's experiences: in both cases, the feeling of abandonment seems to have been a
catalyst for the visions, the visions themselves are overlapping in major
ways, and the personal effects of the visions are about the same (Pascal
moved into an abbey, Sadhu converted to Christianity). Similar cases have
been recorded so many times throughout the ages that it has become a virtual formula: a sincere wish for divine guidance, combined with determi-
nation expressed by anything from resolutely articulated desires to fervent
prayers, sometimes leading to visions.

As was the case with Pascal's encounter with Jesus, there are many
interpretations of Sadhu's vision. In the context of many scientific results,
some of which we have mentioned earlier, specifically in the light of the
strongly substantiated thesis that consciousness can exist incorporeally,
the assessment that what Pascal and Sadhu perceived is sourced in other
'objective' realities is also reasonable, to the extent that anything we perceive is objective.

---

[13] [Stre].

CHAPTER 12

René Descartes

The Philosopher and the Warrior

If you would be a real seeker after truth, it is necessary that at least once in your life you doubt, as far as possible, all things.

RENÉ DESCARTES

René Descartes (1596–1650) is an eminent figure in the history of mathematics; it is likely that he was a mystic too.

12.1 DESCARTES' MATHEMATICS

12.1.1 Coordinate systems

I think that there is no doubt that the most significant mathematical contribution by Descartes is his discovery of the correspondence between (Euclidean) geometry and algebra (which in his times dealt with equations) through the invention of what is now called the Descartes coordinate system.

DOI: 10.1201/9781003282198-12

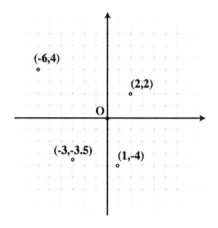

Descartes Coordinate System: points become pairs of numbers and, conversely, pairs of numbers become points. Planar objects become sets of pairs of numbers, and as such they can be studiedanalytically (algebraically)

René Descartes

The introduction of coordinate systems by Descartes set the stage for Newton's and Leibniz's development of infinitesimal calculus.

12.1.2 Euclidean Constructions

A Euclidean construction of an object O in the plane is a construction by means of a ruler (enabling us to draw the line connecting two points) and a compass (enabling us to draw the circle centered at a given point and passing through another point); the object O is then said to be constructible. Descartes showed that for every given positive number d, the number $\sqrt{d}$ is constructible. More precisely, given a line segment of length 1, and another line segment of length d, Descartes found a Euclidean construction of a line segment of length $\sqrt{d}$. This construction is shown in the next figure.

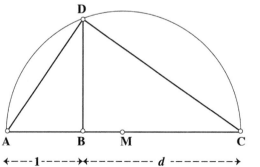

Constructing $\sqrt{d}$.

Descartes constructed BD (as shown in the above figure), and he showed that $BD = \sqrt{d}$. In the figure, AC is the diameter of the shown semicircle and M is the center of that semicircle. By Thales' theorem, the triangle ACD is right-angled. The triangle ABD is right-angled by construction. These two triangles have a common angle at the corner A. Hence, they are similar. The same argument implies that the triangle ACD is also similar to the triangle BCD. Hence the triangles ABD and BCD are similar. It follows that their corresponding sides are proportional. In particular, $\dfrac{AB}{BD} = \dfrac{BD}{BC}$. Since $AB = 1$ and $BC = d$, we get $\dfrac{1}{BD} = \dfrac{BD}{d}$, from where it follows that $(BD)^2 = d$, and so $BD = \sqrt{d}$.

Note: It follows from Gallois theory (early 19th century) that, unlike square roots, cube roots are not constructible. For example, $\sqrt[3]{2}$ is not constructible.

An open problem: the three-dimensional Euclidean tools consist of a three-dimensional compass (enabling us to draw the sphere centered at a given point and passing through another point in the space) and a three-dimensional ruler, or a plane-drawer (enabling us to draw the plane through any three non-colinear points in the space). It is easy to see that everything that can be constructed with the usual Euclidean tools can also be constructed with the three-dimensional Euclidean tools. Can cube roots be constructed with three-dimensional Euclidean tools? In particular, can $\sqrt[3]{2}$ be constructed with three-dimensional Euclidean tools?

12.1.3 Descartes-Euler characteristic

Here are the five regular (Platonic) solids (we will also need them in Section 14.3):

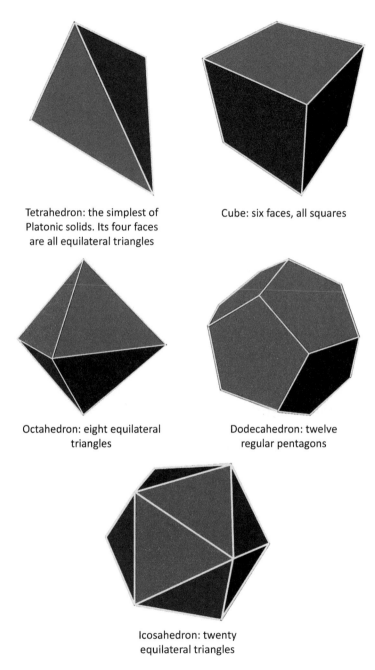

Tetrahedron: the simplest of
Platonic solids. Its four faces
are all equilateral triangles

Cube: six faces, all squares

Octahedron: eight equilateral
triangles

Dodecahedron: twelve
regular pentagons

Icosahedron: twenty
equilateral triangles

The five Platonic solids were mentioned in the never-published secret note-book by Descartes, which was found after his death. This notebook has been lost, but, fortunately, not before it was diligently copied. The timely transcriber was none other than Wilhelm Leibniz!

The secret notebook was written in a coded language. Leibniz might have managed to decode it, but he kept the secret. The story unravels some 350 years after Descartes wrote his notebook. In 1987, the French mathematician and historian Pierre Costabel managed to decipher the secret notebook, and the following observation by Descartes surfaced: the number of faces, minus the number of edges, plus the number of vertices (corners) in Platonic solids is always 2. In symbols: $\#f - \#e + \#v = 2$. This is true for every polyhedron, Platonic or not, whose surface can be *smoothed out* to a sphere. This has become known as the Euler characteristic of polyhedrons, after the Swiss mathematician Leonhard Euler (1707–1783). After the revelation of Descartes' work in 1987, some refer to it as the Descartes-Euler characteristic or formula.

If a polyhedron is not sphere-like, then its Euler characteristic is not 2. For example, if a polyhedron can be smoothed out to a torus (if it is torus-like), then its Euler characteristic $\#f - \#e + \#v$ is 0. So, the Euler characteristic of a polyhedron is a feature of the internal structure of a polyhedron, not of its geometry! We mentioned topology when we covered Brouwer and Poincaré; Descartes' observation centuries earlier may have been the first purely topological result.

12.2 DESCARTES' LIFE

René Descartes was born on March 31, 1596, in north-western France. His mother died a year later while giving birth to her fourth child; the baby died three days later. René was brought up by a governess who was a devout Catholic, so René was raised according to the strict Catholic codes. This notwithstanding, he became remarkably open-minded and unbiased regarding religion-driven politics: when he was in his early 20s and looking for adventures, he volunteered with a Protestant army against the Catholics.

His family was rich; he directly inherited the substantial wealth accumulated by his grandparents, both of whom were doctors. He was financially secure and he didn't need any patrons or salaried positions. He had servants when he needed them, and he was accompanied by a valet (personal attendant) during his entire adult life.

At 11, Descartes entered a local *college*, a Jesuit boarding school; he graduated from there in 1615 when he was 19. Then he moved to Paris, where he studied law and where he got his doctor of law degree a year later. Most of his free time in Paris was spent practicing swordsmanship, and from that time he always carried his sword with him.

He remained in Paris until 1618, living mainly as a bon vivant, filling his days with drinking and gambling; he was a rather successful gambler.

In 1618, he moved to Holland, where he volunteered in the Prince Maurice's Protestant Army against the Catholics, even though, as we have mentioned, Descartes was himself a Catholic. There was a peace treaty, and Prince Maurice's army did not fight. So, in 1619, Descartes went to Germany to join whatever army was accepting volunteers, in the hope of seeing some action.

It was in Germany that he had three 'dreams' that changed his life; more about this further below. On a related subject, while visiting the city of Ulm, Descartes established close contact with the mathematician Johann Faulhaber, whose book *Miracula Arithmetica* (published 1622) influenced Descartes' work. Faulhaber dedicated his work to the 'most enlightened and famous Brothers R.C.' (Rosicrucians), and he may have been a Rosicrucian. He was certainly a mystic and likely also an alchemist. We will come back to these topics in the next section.

In 1620, Descartes was a volunteer in the (Catholic) army of Maximilian of Bavaria, besieging in Prague the (Protestant) forces of Frederick V, the king of Bohemia. Frederick was defeated; he found refuge in Holland. His family included the two-year-old Princess Elizabeth, who some 23 years later, would play a major role in Descartes' life.

In 1621, Descartes quit the service of Maximilian of Bavaria and then traveled through Europe. While in Italy he tried unsuccessfully to meet with Galileo Galilei, whom he greatly admired.

In 1628, he again settled in Holland. He found the Dutch Protestant society more tolerant. Except for relatively short travels, he stayed there for the next 20 years. His first and most important book, *Discourse on the Method* (full title *Discourse on the Method of Reasoning Well and Seeking Truth in the Sciences*), was in French (rather than the commonly used Latin); it was published in 1637 in Holland, and the first printing appeared anonymously. The reason he published it in French was because he intended the book to be understandable to everyone, 'even to women'![1] Later in his life, he changed his unflattering opinion about the intellectual capacity of women.

Descartes' great work on (analytic) geometry, mentioned at the beginning of this chapter, was published as an appendix to *Discourse*.

His other influential book, *Le Monde* (*The World*), was published in 1664, 14 years after his death. In the book, Descartes expressed his support

---

[1] [Kee], p. 22.

for 'heretical' heliocentric theory. So, he delayed publishing it because he was weary of antagonizing the Catholic Church and the Inquisition; he was deeply disturbed by the Inquisition's persecution of Galileo for the same 'offence'.

Descartes once declared[2] that 'a beautiful woman and a good book [] are, of all things in the world, the most difficult to find.' We don't know much about his search for good books, but beautiful women he did find. In 1629, Descartes had more than an ordinary affair with a pretty and intelligent servant named Hélène Jans. He may have secretly married her. In any case, they had a baby girl, Francine, born in July 1635. Francine died from scarlet fever when she was five years old, and the lives of René and Hélène parted.

Descartes' *Discourse on the Method* was widely read at his time by educated people. One of the admirers of his work was Princess Elizabeth, and, since both were in Holland at the time, she wished to meet Descartes. She was 24 years old when they met in 1642; he was 46. They became very close to each other. In his letters, Descartes described her as an angel. Besides being beautiful, she was also very smart; here is a quote from a Descartes' letter to her:[3] 'and I see only Your Highness is a person for whom [both algebra and metaphysics] are equally easy to understand.' Some speculate that they were lovers. They exchanged numerous letters, and Descartes dedicated his *Principles of Philosophy* to her. Like Pascal's Charlotte, Elizabeth fell under Descartes' spell, and she never married, ending her life as a nun.

Elizabeth, Princess Palatine of Bohemia (1618–1680)

[2] [Kee], p. 12.
[3] [Acz], p. 171.

We mention in passing that Elizabeth's younger sister, Duchess Sophia, married Duke Johann Friedrich (John Frederick) of Brunswick, the first patron of Leibniz. As we noted in Chapter 10, Sophia herself was a patron of Leibniz after her husband died.

In 1649, Descartes moved to Sweden, where he was invited by Queen Christina who wanted him as a philosophy teacher. He was brought to Stockholm by the Royal Fleet, which was dispatched to Amsterdam with the sole purpose of picking up Descartes, the only passenger. Five months later, in February 1650, he died from pneumonia. There is suspicion that he was poisoned by the doctor assigned to take care of him. According to this scenario, the doctor resented the lavish attention that Descartes received from the queen, and he was offended by Descartes' steadfast refusal to be treated by him. And there is yet a third explanation, the most extravagant, but perhaps the closest to the truth. We will consider it in the next section.

12.3 DESCARTES' PHILOSOPHY AND MYSTICISM

René Descartes' philosophy was hardly mystic. According to him, the only beings on our planet that have souls are humans; everything else, including plants and animals, are soulless automata. Descartes stated that there exist two things: 'extended substances' (substances that have size, form/figure, and motion) and thinking substances (spirits). The former behave like mechanisms, 'outside of which there is nothing but the soul which manifests to itself its existence through thinking.'[4] The human body is a machine dominated by the brain. The soul (found only in human beings and not in animals) can not be subdivided and is immaterial and immortal. This, in a nutshell, is Descartes' dualism.

Descartes provided a philosophical basis for the subsequent merciless exploitation of our *soulless* planet, including the indiscriminate destruction of flora and fauna.

Descartes produced a syllogistic justification for his claim that existence can be proven through thinking: 'Assume I don't exist; then I could not doubt []. Thus, I must exist.' His famous aphorism, 'Cogito, ergo sum' ('I think, therefore I exist'), is a corollary to this argument. Mystics and mystically inclined philosophers disagree. The mystic philosopher Krishnamurti (mentioned in Chapter 7) claimed a variation of the opposite claim: 'I think therefore I existed', arguing that the act of thinking shifts our consciousness into the past, away from the present, where the

---

[4] Quoted in the introduction of [Lei2].

subject of our thinking originates. On a related subject, Leibniz's monad-ology can be interpreted as postulating the following extension: *I think if and only if I exist.*

Deeper into mystic lore, there arises the thesis stipulating the existence of thinking hallucinations. In Tibetan Buddhism, these thinking halluci-nations are *tulpas*. Rosicrucian variants of *tulpas* are called *homunculi*. Do thinking hallucinations exist? Are we, the humans, *tulpas* of higher enti-ties (or the highest entity, or the Supreme Consciousness)?

Descartes' mysticism, to the extent that it existed, was subdued and secretive. The three 'dreams' that he had as a young man visiting Germany are just about the only explicit report that we have. Descartes regarded them as moments of illumination. Two of the dreams, the first and the third, are just that: dreams; there is nothing mystic about them. The sec-ond dream can be regarded as a case of 'hallucination': he *saw* magnificent sparkles filling the room with light.[5] Since Descartes immediately found a 'satisfactory natural explanation,' the event did not even trigger a reaction that can be called rational mysticism.

So, the thesis that Descartes was mystic rests entirely on the extent of his association with the Rosicrucian Brotherhood (*Fraternitas Rosae Crucis* or The Brotherhood of the Rose Cross). Not much is verifiably known about Descartes' relationship with the Rosicrucian Brotherhood; this is hardly surprising since there is not a single person about whom we can claim with certainty adherence to the Brotherhood. It was really a secretive society.

Who were the Rosicrucians? The main sources of the genesis of the European Rosicrucian Brotherhood are the manifestos *Fama Fraternitatis*, published 1614, and *Confesio Fraternitatis*, appearing soon after, authored by an anonymous author (but allegedly by the same person). According to these two sources, the Brotherhood was established in Europe during the early 15th century by an illuminated sage referred to as 'Our Illustrious Brother and Father C.R.C.' In his youth Father C. R. C. traveled to North Africa and Asia, and while in Arabia he was initiated into ancient occult knowledge. Upon his return to Europe, he gathered seven talented people and established the 'Society of Unknown Philosophers'. These eight were the original members of his society who, after being fully initiated, alleg-edly possessed supernatural powers.

---

[5] [Acz], p. 58.

> They were said to be able to prophecy future events, [] to transform
> Iron, Copper, Lead or Mercury, into Gold, to prepare an Elixir of
> Life, [] by the use of which they could preserve their youth and
> manhood [!]; and moreover it was believed that they could com-
> mand the Elemental Spirits of Nature, and knew the secrets of the
> Philosopher's Stone, a substance which rendered him who pro-
> cessed it all-powerful, immortal, and supremely wise.[6]

In short, they were realized mystics: upon returning back from their
mystic unions into corporeal existence, they brought back with them the
exalted cosmic consciousness.

We mentioned that in 2019 Descartes met with the German mathematician Johann Faulhaber, who was likely associated with the Rosicrucian
order, and who may have introduced Rosicrucianism to him. In any case,
when Descartes returned to Paris in 1621, it was widely rumored that he
had joined the Brotherhood while in Germany. Since the Catholic Church
forbade any association with Rosicrucians, Descartes was in a rather pre-
carious situation; there were at the time plenty of religious zealots ready to
lynch any alleged heretics. So, prudently enough, he publicly declared his
opposition to Rosicrucianism and denied being associated with it.

If he lied, then there was another reason for doing so: the Brotherhood
was strictly secretive: under no circumstances were its members, called
'Invisibles', allowed to reveal their allegiance.

So then, had Descartes been initiated to (some of) the secrets of the
Rosicrucian society, and, if so, had he accepted the genuineness of these
(occult or esoteric) secrets? The answers to both of these questions seem
to be affirmative.

Descartes' secret notebook was mentioned in Section 12.1.3; recall
that the original was lost and that we only have Leibniz's transcription.
The notebook, which was never intended to be seen by anyone else but
Descartes himself, is awash with Rosicrucian symbols. On top of that, the
notebook contains an explicit reference to the Rosicrucian Brotherhood:
the *Preambles* of the notebook contains the following dedication: *Offered,
once again, to the erudite scholars of the entire world, and especially to
G.F.R.C.*, which stands for *Germanicus Fraternitas Rosae Crucis*. Lest this
be dismissed as a product of Leibniz's intervention, there is another similar dedication: in 1620, Descartes wrote a mathematical treatise entitled

---

[6] From *Secret Symbols of the Rosicrucians* by Dr. Franz Hartmann, quoted in [Hall], p. 447.

Polybiicosmopolitani Thesaurus Mathematicus. All that survives today is the title and the fact that it was dedicated to the Rosicrucians. These two dedications clearly indicate Descartes' close association with the Rosicrucian Brotherhood.

Finally, there is a very intriguing story regarding Descartes' death, with some supporting evidence of its genuineness. The French scholar Pierre Daniel Huet (1630–1721), in his memoirs published in 1692, wrote that he felt Descartes did not die in 1650 but faked his funeral, presumably in order to become 'invisible' and devote the remainder of his life to attaining spiritual illumination. Huet and other writers cited letters written in 1652 and 1656 between Descartes and Queen Christina, supposedly published by Charles Adam and Paul Tannery in *Oeuvres de Descartes* (*Works of Descartes*).[7] If true, this would make Queen Christina a co-conspirator in the fake funeral.

An obvious objection to this conspiracy is Descartes' body: it must have been seen by those attending his funeral. Perhaps not: somewhat incongruently with the royal treatment that Descartes received in Sweden, his (or someone else's) body was in a coffin that was buried in a humble cemetery for orphans and Catholics during a funeral attended by very few people. Descartes' remains were exhumed and reburied in Paris 16 years later. The body was headless (!), which is what one would expect in the case of a fake funeral.

All and all, there seems to be enough indication supporting the genuineness of the thesis that Descartes was associated with the Rosicrucian Brotherhood.

---

[7] *Awakened Attitude* by Gary L. Stewart; see https://www.crcsite.org/rosicrucian-library/contemporary-writings/rene-des-cartes/.

Cardano

The Psychic Mystic

There is no reality except the one contained within us.

HERMANN HESSE: *DEMIAN*

Pascal lamented, 'we cannot think of two things at the same time'. About a century earlier Cardano claimed that he could! Girolamo Cardano is the main subject of this chapter.

13.1 THE STORY OF THE CARDANO'S FORMULA

The formula for solving cubic equations is rarely given in modern textbooks; we provide it here, together with a short history of its discovery.

Consider the equation $ax^3 + bx^2 + cx + d = 0$, where $a \neq 0$, and $b, c,$ and d are constants. We may assume $a = 1$ (otherwise divide the whole equation by a) so that the equation is $x^3 + bx^2 + cx + d = 0$.

Set $e = \dfrac{3c - b^2}{9}$, $f = \dfrac{9bc - 27d - 2b^3}{54}$, and $g = \sqrt[3]{f + \sqrt{e^3 + f^2}}$, $h = \sqrt[3]{f - \sqrt{e^3 + f^2}}$.

DOI: 10.1201/9781003282198-13

Then the solutions of the equation are $x_1 = g + h - \dfrac{b}{3}$, $x_2 = -\dfrac{g+h}{2} - \dfrac{b}{3} + \dfrac{i\sqrt{3}}{2}(g-h)$,

and $x_3 = -\dfrac{g+h}{2} - \dfrac{b}{3} - \dfrac{i\sqrt{3}}{2}(g-h)$, where $i = \sqrt{-1}$. This is now known as

Cardano's formula.[1]

We must note that the modern notation we have used greatly simplifies our exposition. In Cardano's times even the equality symbol $=$ had not yet been invented; it was introduced by Michael Recorde in 1557 and later popularized by Leibniz.

Let's apply the formula to $x^3 - x^2 + x - 1 = 0$; we have, in turn:

$$b = -1,\ c = 1,\ d = -1,\ e = \frac{2}{9},\ f = \frac{10}{27},\ g = \sqrt[3]{\frac{10}{27} + \frac{2}{3\sqrt{3}}},\ \text{and}\ h = \sqrt[3]{\frac{10}{27} - \frac{2}{3\sqrt{3}}}.$$

By Cardano's formula we have that $x_1 = \sqrt[3]{\dfrac{10}{27} + \dfrac{2}{3\sqrt{3}}} + \sqrt[3]{\dfrac{10}{27} - \dfrac{2}{3\sqrt{3}}} + \dfrac{1}{3}$,

which simplifies to $x_1 = 1$,

$$x_2 = -\frac{1}{2}\left(\sqrt[3]{\frac{10}{27} + \frac{2}{3\sqrt{3}}} + \sqrt[3]{\frac{10}{27} - \frac{2}{3\sqrt{3}}}\right) + \frac{1}{3} + \frac{i\sqrt{3}}{2}\left(\sqrt[3]{\frac{10}{27} + \frac{2}{3\sqrt{3}}} - \sqrt[3]{\frac{10}{27} - \frac{2}{3\sqrt{3}}}\right),$$

which simplifies to $x_2 = i$, and

$$x_3 = -\frac{1}{2}\left(\sqrt[3]{\frac{10}{27} + \frac{2}{3\sqrt{3}}} + \sqrt[3]{\frac{10}{27} - \frac{2}{3\sqrt{3}}}\right) + \frac{1}{3} - \frac{i\sqrt{3}}{2}\left(\sqrt[3]{\frac{10}{27} + \frac{2}{3\sqrt{3}}} - \sqrt[3]{\frac{10}{27} - \frac{2}{3\sqrt{3}}}\right),$$

which simplifies to $x_3 = -i$. All of this can be easily confirmed since

$x^3 - x^2 + x - 1$ happens to be $(x-1)(x^2+1)$.

In the last example, we have encountered the complex number i defined[2] by $i = \sqrt{-1}$. It is interesting to note that in order to find the real (ordinary) solutions of a cubic equation by means of Cardano's formula, it is necessary to manipulate complex ('imaginary') numbers. Cardano was the first who performed some simple operations with complex numbers, so the invention of complex numbers $a + bi$ is sometimes attributed to him.

However, and curiously, Cardano's formula itself is not due only to Cardano! A number of Cardano's contemporaries contributed. The rather interesting story of Cardano's formula is given in details in Ore's book *Cardano, the Gambling Scholar*. We summarize it here.

---

[1] There are some fine points in the formula related to the non-uniqueness of the roots (radicals); we will not provide details.

[2] In fact, there are two square roots of −1: i and $-i$.

The plot starts with Niccolò Fontana (1499–1557), better known as Tartaglia (the stutterer). In 1535 Tartaglia had a public dispute challenge with Antonio Fiore. These polemics, a kind of academic battles of gladiators, were taken very seriously, and they often involved large sums of money. On top of that, these duels were adjudicated by people of high standing, and the outcomes of the disputes were taken into account when the participants applied for positions of merit. So, the stakes were rather high. The match between Tartaglia and Fiore consisted of 30 questions that both participants assigned to each other. The loser was to pay for 30 banquets for the winner and his friends. All of the 30 questions Fiore assigned to Tartaglia involved solving certain cubic equations of type $x^3 + cx + d = 0$. Tartaglia managed to find a general method for solving <u>such</u> equations, solved all 30 problems, and was declared the winner.

Four years later, in 1539, Cardano was about to finish his book *The Practice of Arithmetic, Geometry, and Algebra*, and approached Tartaglia in order to procure permission from him to publish his method for solving cubic equations. Tartaglia refused to divulge his methods. He did not change his mind even after Cardano offered him a separate chapter under Tartaglia's name. However, after years of persuasion, Tartaglia agreed to reveal his method but only after Cardano took a solemn oath that he would under no circumstances publish the discovery.

Soon after Cardano received Tartaglia's solution he managed to generalize and find a method for solving general cubic equations $x^3 + bx^2 + cx + d = 0$; he did it by cleverly reducing them to the types of equations considered by Tartaglia. However, Tartaglia steadfastly refused to allow publication.

In 1543, Cardano and his young and talented disciple Ludovico Ferrari journeyed to Bologna to examine the posthumous papers of Scipione del Ferro, who had been a professor of mathematics there. They confirmed the rumors that they had heard, that del Ferro had found a correct method for solving equations of type $x^3 + cx + d = 0$ years before Tartaglia. This freed Cardano's hand, and in 1545, he published the book *Ars Magna* that included a complete method for solving $x^3 + bx^2 + cx + d = 0$ (as well solutions for $x^4 + ax^3 + bx^2 + cx + d = 0$, due to his student Ludovico Ferrari). In the book, Cardano generously acknowledges Tartaglia's contributions, even though Tartaglia had merely rediscovered del Ferro's result.

13.2 CARDANO: A SHORT BIOGRAPHY

Gerolamo Cardano

Girolamo Cardano (Italian), a.k.a. Jerome Cardan (English), a.k.a. Hieronimus Cardanus (Lattin) (we will use his original, Italian name), was born in 1501 in Padua. His father Fazio, who will also feature later in our narrative, was a lawyer in Milan. Fazio had wide interests, and at one time he was appointed a public lecturer in geometry; Leonardo da Vinci noted that he had consulted him in geometry several times. Fazio met Gerolamo's mother Chiara late in his life. At that time, Chiara was a widow with three children, all of whom died during a plague in Milan. It has been said that Chiara tried to abort Gerolamo, perhaps because the baby was conceived out of wedlock. Chiara was in labor for three days while giving birth to Girolamo. Fazio married Chiara when Gerolamo was a teenager, and he died soon after.

Gerolamo was a sickly child but managed to pull through. At 19, he was accepted at the university in Padua to study medicine, and five years later, in 1925, he was awarded the degree of Doctor of Medicine. The same year he applied for admission at the College of Physicians of Milan, but was denied, ostensibly because of 'bastardy', but likely because he already had antagonized some influential people in Milan. Between 1526 and 1532, he was a country doctor in a village near Padua called Sacco. In his biography, he described the time when he had lived in Sacco as the happiest period in his life.

When he was a young man, Girolamo had to put up with an infliction that he attributed to his difficult birth. He writes:[3] 'Since Jupiter was in the

---

[3] [Car], p. 4.

ascendant and Venus ruled the horoscope, I was not maimed [during his mother's long labor], save in genitals, so that from my twenty-first to my thirty-first year I was unable to lie with women.' This problem was solved in Sacco after he married. Soon after, his family increased by two children: a boy and a girl. The second son was born seven years later.

In 1532, he moved to a little town near Milan but was forced to move again, to Milan, where, for a while, he and his family lived in a poorhouse shelter. Eventually, he was appointed to the position of public lecturer in mathematics, a position that was once held by his father. In 1539, he was finally admitted to the College of Physicians of Milan, and soon after he become the most prominent physician of Milan. His reputation was fast rising, and in a few years, he was considered to be one of the two best physicians in Europe. His expertise was sought by many royals and church dignitaries, including the king of Denmark and the pope. In 1952, he traveled to Scotland where he treated and cured the Archbishop of Scotland, John Hamilton, of his asthma. The jubilant archbishop, who remained almost asthma-free all his life, awarded Cardano the substantial sum of 1,800 gold crowns and a heavy gold chain.

Not surprisingly, Cardano considered himself first and foremost a physician. However, he was multitalented and active in many other disciplines, not the least of which was mathematics.

He had great respect for mathematical knowledge; he wrote in his book *De arcanis aeternitatis*: 'Only in mathematics does the human brain acquire absolutely valid knowledge.'[4]

His first two mathematical textbooks were published in 1539, and they were followed by many others. In his autobiography *The Book of My Life* (*De Vita Propria Liber*) he claimed, somewhat outlandishly, that he had solved 40,000 major and 200,000 minor problems or questions.

He was very versatile and wrote many books on wildly varying subjects: mathematics, death, immortality, chess, etc. Here are a few titles: *On Poisons, On Dreams, On imaginary expressions* [Complex numbers], *On Music, On the Urine, On Games of Chance*! During his lifetime Cardano published 131 books, with some of the remaining manuscripts – a total of 111 – being published after he died. His *Opera Omnia* in Latin, published in 1662, consists of ten large volumes containing most of his work.

As we saw above, he is given credit for the discovery of complex numbers and for his contributions to solving cubic equations. On top of this, he wrote the gambler's handbook *Ludo Aleae* (*The Book on Games of Chance*),

---

[4] [Car2], p. 11.

which contains the basics of probability theory. Some consider it to be the first text on the theory of probability. (The book is given in full as the last chapter of Ore's book.)

His personal reason for writing such a book was rather straightforward: Cardano was a compulsive gambler for long periods of his life. The following dramatic and colorful story comes from his autobiography. Once when he was 25, he gambled for two days in the 'house of a cheat' in Venetia. When he discovered that the cards had been marked, he 'drew [his] dagger and wounded [his opponent] in the face'. Then he grabbed all the money, his and the cheat's, attacked the servants in the house, and somehow managed to escape, only to slip and fall into the sea. He did not know how to swim, but, luckily, he was pulled up by the passengers of a passing boat. The man who he had wounded was in the boat, with his face bandaged. This man was in a conciliatory mood, and he offered Cardano clothing, so, somewhat melodramatically, all ended well.

Cardano was also a chess addict! Here we quote him from his autobiography:[5]

> It was in the summer of the year 1542 when I fell into the habit of going every day to the house of Antonio Vimercati, a noble in the town, for the purpose of playing chess. [] But through this habit I had fallen so low, that for two years and some months I neglected my medical practice, my other incomes, my reputation, my studies.[6]

In 1546, Cardano's wife died. At the time, his children were twelve, ten, and three. The oldest son, Giambatista, seemed to have inherited his father's (and grandfather's) talents and was Cardano's favorite. When Giambatista was 22, he married a woman of dubious reputation without giving notice to his father. Their marriage turned out to be rather inharmonious, and during one violent quarrel, Giambatista's wife ridiculed and insulted him by claiming that he was not the father of their three children. Giambatista was beside himself with anger and poisoned his wife. He was jailed, sentenced to death, tortured, and guillotined. He was 26 at that time.

The horrible destiny of his child broke Cardano's heart. He became desperately distressed and could not stop thinking about the tragedy. So, after

---

[5] Also in [Ore], p. 109.
[6] Here I cannot help but feel strong empathy for the great men: unrestrained chess-playing is exactly what I have been doing in a period of my own life.

deliberating upon his misfortune, he decided to pray for his own death. His morbid plans were thwarted by an 'absolutely supernatural' event that we will recount in the next section.

Cardano described himself as being of medium height, with very small feet, a narrow chest, thin arms, a long neck, a heavy lower lip, a wart over one eye, a beard parted in two points, and a loud voice that irritated even his friends. He had many diseases (kidney trouble, heart palpitations). When he wrote his autobiography he was 68, and at that time he had 14 good teeth and one bad! He notes succinctly:[7] 'I am a man bereft of bodily strength, with few friends, small means, and many enemies.'

His personality was an exuberant combination of many contradictory threads (quoting Cardano):[8]

> Nature has made me capable in all manual work, it has given me the spirit of a philosopher, and the ability in the sciences, taste and good manners, voluptuousness, gaiety, it has made me pious, faithful, fond of wisdom, meditative, inventive, courageous, fond of learning and teaching, eager to equal the best, to discover new things and make independent progress, of modest character, and student of medicine, interested in curiosity in discoveries, cunning, crafty, sarcastic, an initiate in the mysterious lore, industrious, diligent, ingenious, leaving only from day to day, impertinent, contemptuous of religion, grudging, envious, sad, treacherous, magician and sorcerer, miserable, hateful, lascivious, solitary, disagreeable, rude, divinator, envious, obscene, lying, obsequious, fond of the prattle of old men, changeable, irresolute, indecent, fond of woman, quarrelsome, and because of the conflict between my nature and soul I am not understood even by those I associate most frequently.

Cardano died in 1578.

13.3 THE WILD MYSTICISM OF CARDANO

Cardano the elder, Fazio, who, as we recall, was also proficient in the mathematics of his times, had an extraordinary experience long before Girolamo was born. The account below comes from Cardano's book *De Subtilitate*:

> Here I will add a story which is more wonderful than all the rest, and which I have heard from my father, Facius Cardan (who

---

[7] [Car], p. 6.
[8] [Ore], p. 25.

confessed that he had had a familiar spirit for nearly thirty years) recount not once but many times. Finally, I searched for his record of this event, and I found that which I had so often heard, committed to writing and to memory as follows.

August 13, 1491. When I had completed the customary rites, at about twentieth hour of the day, seven men dully appeared to me clothed in silken garments, resembling Greek togas, and wearing, as it were, shining shoes. []

Nevertheless, all were not dressed in this fashion, but only two who seemed to be of nobler rank than the others. [] When asked who they were, they said that they were men composed, as it were, of air, and subject to birth and death. It was true that their lives were much longer than ours, and might even reach to three hundred years duration. Questioned on the immortality of our soul, they affirmed that nothing survives that is peculiar to the individual. They said that they themselves were more closely related to the gods than mankind, but were yet separated from them by an almost immeasurable distance. [] They said that no hidden things were unknown to them, neither books nor treasures, and that the basest [of lowest rank] of them were the guardian spirits of the noblest of men, just as men of low degree are the trainers of good dogs and horses. They have such exceedingly subfile bodies that they can do us neither good nor harm, save through apparitions and terrors or by conveying knowledge. The shorter of the two leaders had three hundred disciples in a public academy, and the other, two hundred. []

[] They remained with my father for over three hours. But when he questioned them as to the cause of the universe they were not agreed. The tallest of them denied that God made the world from eternity. On the contrary, the other added that God created it from moment to moment, so that should He desist for an instant the world would perish.

So, Fazio had a meeting with his (hallucinated or not) familiar spirits. Girolamo himself experienced in his lifetime, especially while he was a child, a profusion of apparitions. Between the fourth and seventh years of his life, between the start of the day and until the 'third hour of the day',

Cardano used to have 'visions' consisting of[9] 'images of airy nothingness of body'. He writes in his autobiography:[10]

> These images arose from the lower right-hand corner of the bed, and, moving upward in a semicircle, gently descended on the left and straightway disappeared. The images of castles, of houses, animals, of horses and riders, of plants and trees, of musical instruments, [], of men of divers costumes [?] and varied dress; images of flute-players, with their pipes as it were, ready to play, but no voice or sound was heard. [] I could see a veritable chaos of innumerable objects rushing dizzily along en masse, without confusion among themselves, yet with terrific speed. These images were, moreover, transparent, [] yet so dense as to be impenetrable to the eye. [] Even flowers of many a variety, and four-footed creatures, and divers birds appeared in my vision.

These are hypnopompic (visual) hallucinations, as opposed to the more prevalent hypnagogic hallucinations happening during the transition from waking to sleeping consciousness. In the next section, we will say more about such and other hallucinations, and about their ubiquity. We will only mention that we[11] do not accept the prevailing orthodox-medical view of hallucinations according to which they are necessarily symptoms of mental or physical pathology.

Cardano also experienced auditory verbal hallucinations; in fact, he recorded some conversations with discarnate entities, spirits, or, as his father Fazio said, with men composed of air, in writing. We alluded to one such conversation earlier: we mentioned that one night Cardano was praying for his death since he could not stop thinking of his son's tragic demise. That night he had a vision – or an encounter with some entity in the darkness. The *hallucinated spirit* instructed him:[12] 'place the gem which you have on a chain about your neck in your mouth, and as long as you keep it there, you will not remember your son.' In the morning he decided to give it a try: 'I placed the gem in my mouth, and lo, a thing occurred beyond all belief; for at once every memory of my son was lost in oblivion!' He could not have the gem in his mouth while eating or lecturing, and at this

---

[9] [Car], p. 147–148.
[10] Ibid.
[11] 'we' comprises me and Cardano.
[12] [Car], p. 208.

time he was tormented again. He mentioned though that these miraculous changes from one condition to another happened instantaneously the moment he placed the gem in his mouth or took it out from his mouth!

For the orthodoxy, Cardano's experience was anything between a fantasy and a baseless hallucination followed by a placebo effect. Explaining it as a deliberate lie should be excluded: we recall that Cardano wrote his autobiography with (at times) self-defeating sincerity and honesty. However one interprets these events, he himself believed them to have been completely authentic miracles.

Cardano was well aware of the disbelief that he would encounter; he bemoans:[13] 'I know certain men who [] scoff and incite derision in regard to any such apparently supernatural occurrence.'

We recall that Ramanujan figured out from studying the lines on his palm that he would die before he was 35. Hinduism and Buddhism considered palmistry a discipline of basic education. The Buddhist monk Bhavaviveka, who lived around the 6th century, wrote:[14] 'one should thoroughly study the subjects which are well known [!] in the mundane world, i.e., grammar, palmistry [!], enumeration, alchemy [!], medical science, arithmetic, charms [!], spells [!], etc.' Cardano agreed wholeheartedly and practiced just about everything Bhavaviveka mentioned. He carried various amulets and precious stones, that he believed protected himself against harm. And, in his 13 books on divination, he wrote about palmistry.

Cardano also diligently studied astrology, or, as he wrote,[15] 'the branch of astrology which teaches the revealing of the future.' He was considered an expert in astrology, and his astrological readings were sought by many dignitaries. We do not have the statistics, but it is reasonable to postulate that his advice was deemed valuable to his clients most of the time. However, it is known that he once failed spectacularly. On his way back from Scotland he was a guest of King Edward the Sixth, the 16-year-old boy-king of England. The young king requested an astrological reading of his future life from Cardano. Cardano predicted a long life for the king and unhesitatingly supplied some details of this long life. The king died soon thereafter.

---

[13] Ibid, p. 212.
[14] [Iid], p. 69.
[15] [18Car], p. 169.

Cardano could put himself into an altered state of consciousness, or 'ecstasy', at will, and then have a kind of light OBE (out-of-body experience). He writes:[16]

> When I go into trance I have near my heart a feeling as though the spirit detached itself from the body, and this separation extends to all the body, especially the head and neck. After that I have no longer the idea of any sensation, except of feeling myself outside my body.

Further, he relates that in that state he was insensible to any pain. Cardano wrote the following about ecstasy:[17] 'An Ecstasy is a kind of medium between sleeping and waking, as sleep is a kind of middle state between life and death. Things seen in an Ecstasy are more clear, and far more evident.' Intensely enhanced (incorporeal) perception is one of the main features of OBEs.

Cardano experienced vibrations whenever he was the topic of discussion somewhere, possibly far away from him. He writes:[18] 'Very often when the discussion about me has taken place in the same city, it has happened that the vibration has scarcely ceased before the messenger has appeared who addressed me in the name of my detractors'. And in 1534 he began experiencing precognitive dreams:[19] 'I began to see in my dreams the events shortly to come to pass.'

His 'intuitive flashes of direct knowledge' had different manifestations. For example, he claimed that even though he[20] 'never learned the Greek, French or Spanish languages, but the ability to employ these have been given to me; how, I cannot explain.'

All in all, Cardano's life was flooded by a formidable array of *paranormal* wonders. He even saw (what we now call) UFOs – two times:[21]

> When I was a lad I saw a star like Venus shining so high in heaven at the twenty-second hour of the day that it could be seen from our whole city. Thereafter, I saw three suns in the year 1531 and all casting their broad beams across the eastern sky. [] The spectacle lasted about three hours altogether.

---

[16] [Bla], p. 31; the original source is not given.
[17] Quoted in [Sum], p. 93.
[18] [Car], p. 164.
[19] Ibid.
[20] Ibid, p. 168.
[21] Ibid, p. 192.

It is interesting to note that in the year 1494 a similar phenomenon – 'three suns in the night skies' – was observed by many people in the vicinity of Naples.

13.4 HALLUCINATIONS, PRECOGNITION, TIME, AND PLACEBO REALITIES

Cardano was convinced that the hidden connections in nature could only be grasped by intuition. Here, he was ahead of his time for some 450 years. There are strictly controlled experiments proving with statistical certainty that artistically inclined people, for example, musicians, who rely more on intuition (right brain dominance) score substantially above chance in experiments related to extra-sensory perception, compared to people who rely mostly on their analytical skills (left brain dominance). For example, in the *ganzfeld* experiments[22] done by Braud, Honorton, Parker, and others (published in a paper by Honorton), the sender would concentrate on an image, randomly chosen among four, and the target-receiver (in a distant room with red flooding light, with ping-pong half-balls glued to the eyes, and with white noise in headphones) would then choose one of these four images as being the closest to what he perceived during the experiment. The odds of success are 25%. There was a total of 28 studies, and the cumulative success average was 35%. The odds of that happening by chance is in the order of 1:10 billion. The success rate for the musicians as targets was a whopping 50%.

Since Cardano was a great hallucinator, we now pay attention to hallucinations. As we have mentioned earlier, hallucinations are standardly defined – somewhat vaguely – as percepts that are not sourced in the (material) reality. There are many types of hallucinations; we have mentioned hypnagogic (on the boundary from waking to sleeping) and hypnopompic (on the boundary from sleeping to waking) hallucinations. Most of the Cardano hallucinations were of one of these two types. According to many polls, the majority of people experience at least one occurrence of hallucination in their lifetimes. As a consequence of this ubiquity, they are not automatically considered symptoms of (mental or physical) pathology; the diagnosis is usually deferred to the next step and depends on the attitude of the experiencers following the hallucination. If you are not 'fooled' and if you agree with the orthodoxy that the event is a brain product that is perceived as an external projection, you are sane. If you think it is 'real' and if you consider seriously that the event is sourced out of your mind,

---

[22] [May].

then you are at least proto-schizophrenic. Explaining hallucinations as being necessarily produced by our brains does not account for group hallucinations (when an 'apparition' is perceived simultaneously by at least two people – there exist cases of apparitions perceived at the same time by hundreds of people) and for apparitions that have veridical aspects (there are, for example, hundreds of authenticated published cases of a hallucinator seeing an apparition of a person that had died at about the same time). We will, thereby, disagree most categorically with the opinion of some historians that Cardano's sanity is questionable. Elaborating further necessitates going deeper into anecdotal evidence; we have decided to avoid anecdotes.

We mentioned that Cardano often foresaw events that eventually 'came to pass.' His precognition was sometimes in the form of premonition, often in dreams. Sometimes this foreknowledge manifested in rather unusual ways: he had stigmata of a cross lasting for a couple of months before his son was beheaded, disappearing a day after the execution.

Precognition has been tested in various labs many times and has been established with statistical certainty as a genuine phenomenon. For example,[23] the simple test consisting of subjects guessing the number of objects that were randomly chosen *after* their guess was performed nearly two million times between 1935 and 1987. The overall results were positive with odds against chance around 10^{-25}.

We will mention only two more series of experiments in precognition. In one of the Jahn-Dunne experiments[24] with significantly positive results, the receiver 'got' images from the sender up to three days *before* it was selected by a computer and 'sent' by the sender. In the second batch of experiments, the Hungarian researcher Zoltan Vassy administered painful electric shocks as stimuli to be precognized by recipients, and, not surprisingly, obtained very strong positive results.[25]

I must admit I know next to nothing about astrology, and so there is the temptation to dismiss it off hand, thus automatically assuming a posture of self-congratulatory objectivity. Indeed, the idea that our lives are somehow related to the mutual position of the planets seems extremely preposterous. On the opposite side of the argument, one of the most basic experiential mystic tenets is that everything is interconnected and that all consciousnesses, from atomic to human to planetary to the pervasive supreme consciousness, permeate one another. So, from the vantage point of mysticism, astrology is a reasonable endeavor. Moreover – oh heresy,

---

[23] [Rad2], p. 149.
[24] [Dos], p. 199.
[25] [Targ2], p. 87.

oh heresy! – there exists some scientific support! Check, for example, the book entitled *Astrology; The Evidence of Science* (by Percy Seymour[26]).

Here is a modest experiment: the psychiatrist Eugene Jonais, based on clues found in an ancient book of Vedic astrology, predicted the sex of the fetuses of his clients 17 consecutive times.[27] This happened before ultrasound scanners were in use. The probability of this happening at random is, of course, 2^{-17}, which is 1 out of 131,072 – not astronomical, but rather significant.

The most extensive scientific analysis of astrology seems to have been done by the French psychologist Michel Gauquelin, who found a significant statistical correlation between the positions of the planets at birth and people's professions;[28] see his book *The Scientific Basis for Astrology*.

If defending astrology is a heresy from the perspective of materialistic science, then defending palmistry is probably a scientific lunacy. In a way, that is exactly what I am now going to do, even though I couldn't care less about divinations such as palmistry!

How do we interpret Ramanujan's conviction that the lines on his palm foreboded his early passing? The most obvious common interpretation is that it was a coincidence. Next in line is explaining it through a kind of reversed placebo effect: Ramanujan convinced himself that he was going to die young and, consciously or subconsciously, steered his life in that direction.

There is yet another interpretation, that, at first sight, may seem outlandish: Ramanujan, through his beliefs subconsciously (overconsciously?) *choose* a stream of time that was to be an aspect of the reality of his life. The thesis is that true belief in a palmistry reading steers the life of the believer in the direction of that reading.

Perhaps surprisingly, the theory according to which perceived reality, ostensibly independent from the experiencers, is shaped by our strong beliefs, is more than an unsubstantiated mystical speculation. Consider the well-known and statistically supported sheep-goat effect. In 1943, the psychologist Gertrude Schmeidler predicted,[29] and confirmed statistically in card-guessing tests, that those who accept the idea of extrasensory (psi) perception (the sheep) will do better at psi tests than those who reject it (the goats). In 1993, psychologist Tony Lawrence confirmed Schmeidler's hypothesis.[30] He conducted a meta-analysis of the sheep-goat

---

[26] [Sey].

[27] [Gro2], page 336.

[28] The so-called debunkers, most of whom subscribe to the 'that-couldn't-be' axiom, do not interest me.

[29] [89Pla], p. 32.

[30] [Rad], p. 89.

forced-choice tests (guessing one of a finite number of objects that are known to the guesser) conducted between 1943 and 1993. He found 73 publications reported by 37 different investigators, involving over 685,000 guesses produced by 4,500 participants. The overall results showed that the believers (sheep) performed better than disbelievers (goats) with odds against chance greater than a trillion to one.

The sheep-goat effect, as it seems, applies not only to the subjects of the test but to the experimenters as well: the results of the tests performed by scientists who believed in a specific outcome (the sheep) consistently and with significant statistical certainty yielded more positive results compared to the results of the test performed under identical conditions but carried out by scientists who did not believe in such an outcome (the goats). This is harder to substantiate further since there is only a small number of studies where the experimenters are at the same time subjects. However, there is the case of Marilyn Schlitz, a successful psi investigator, and the skeptic Richard Wiseman.[31] They carried out separate experiments in remote viewing, three by Schlitz and two by Wiseman; the experiments were conducted in the same location, using the same equipment, and with subjects drawn from the same pool. All three of Schlitz's experiments showed statistically significant evidence of a psi effect, while in both of Wiseman's experiments, the results were at chance.

Admittedly, the above experiment is insufficient to draw conclusions; there are many reasonable interpretations. One such explanation is the claim that the beliefs of the experimenters strongly influence the, ostensibly independent, (psi-)nature of reality: 'sheep' witness worlds with an abundance of psi phenomena, and goats live in mundane worlds. The probability that perceived reality will match our beliefs seems to increase with the intensity of these beliefs. In a way, perceived reality is a placebo reality.

So then, perhaps the same is true for palmistry: perhaps the probability of the realization of a reading of the palm lines increases with the strength of the belief of those involved in the reading: believers increase, perhaps by just a little bit, the probability that the prediction will come true, and skeptics decrease that probability. Perhaps societies where palmistry is considered viable, like the ancient Vedic Indian world, influence the probability of realizing the streams of time where the corresponding divinations are more regularly on target.

We live in a world shaped by the beliefs of skeptics.

---

[31] [Rao], p. 19.

CHAPTER 14

A Gallery of Mystics and Mathematicians

Second Digest

Call no man happy until he is dead.

OVID

14.1 EMANUEL SWEDENBORG: THE INTREPID OBE TRAVELER

Emanuel Swedenborg (1688–1772) was a minor mathematician and a major mystic. He could have left deeper a trace in the history of mathematics; however, like Cardano, Pascal, Leibniz, and Newton, he was multitalented, and, ultimately, the roads of destiny led him in the direction of realizing some other of his talents.

DOI: 10.1201/9781003282198-14

Emanuel Swedenborg

Emanuel was born in Stockholm in 1688 to an affluent family: both of his grandfathers were successful mining businessmen, and his father, Jesper, was an eminent bishop. Between 1692 and 1702, his family settled in Upsala (or Uppsala) near Stockholm, which was where he started his college studies.

He started his adult life with mathematics as his main interest. In the year 1710, when he was 22, he traveled to England intending to study mathematics and astronomy. Crossing the open seas between Sweden and England was not a routine affair those days. The vessel in which he sailed almost sank in the rough waters off the English coast, and soon after it was boarded by pirates. In the ensuing confusion, the ship was fired upon by English coast guards. Finally, Swedenborg was almost hanged after landing in England because he violated the quarantine rules established as a precaution against a plague that ravaged Sweden at the time.

He stayed in London for two years. He wrote in a letter: 'I study Newton daily and am very anxious to see and hear him.' It seems that that never happened, which is not surprising given what we know about Newton's secluded life. However, he did get in touch with other English mathematicians and learned men. He writes[1] in April 1711: 'I visit daily the best mathematicians here in town.'

After England, Swedenborg traveled to the Netherlands, France (Paris), Italy, and Germany, continuing his studies and expanding his knowledge.

---

[1] [Tro], p. 19.

He returned to Sweden in 1715. His practical father, who did not look favorably on a career in mathematics for his son, through his influence secured for the young Swedenborg an appointment as 'Extraordinary Advisor of the Board of Mines', the position that he eventually became sufficiently fond of, since it allowed him relative intellectual freedom.

At about that time, Swedenborg got engaged to the daughter of an older friend. When she ended up marrying someone else, his friend then promised him his younger daughter, and she was formally betrothed to him. She, like her sister, was also adamantly opposed to the engagement with Swedenborg. Her father was very fond of Swedenborg and insisted on their marriage. However, the teenage girl stood fast in her opposition; her prince charming was not exactly epitomized in Swedenborg, who was a shy stutterer. Upon seeing her grief, Swedenborg gallantly released her from the contract. At the same time, he vowed never again to have any affectionate relationship with a woman. So, his romantic life resembled Pascal's. We cynically observe that had these two great people had happier romantic lives, we would likely have been deprived of Pascal's *Pensées* and Swedenborg's accounts of his amazing spiritual travels.

Swedenborg's original last name was Swedborg. In 1719, by a decree of the queen, Swedenborg's family was 'ennobled', and the last name was changed to Swedenborg, the name we have consistently used, independent of the chronology.

Meanwhile, Swedenborg continued his work in mathematics and published the book *Rules of Algebra*. In 1717, he was offered the position of Professor in Mathematics and Astronomy at the University of Upsala; he declined it.

He also published scientific books in chemistry and physics, a treatise on the construction of docks and dykes, works on geology and mineralogy, and notes on some practical inventions related to mining. Later, when he was in his 40s, Swedenborg published a few influential scientific monographs, one of which was the well-received work entitled *Opera Philosophica Mineralia*. He was one of the first elected members of the Royal Academy of Sciences of Sweden. In 1734, he was invited by the Academy of Sciences of St. Petersburg to become a correspondence member. So, at the peak of his career, Swedenborg was above all a distinguished scientist.

When he was in his late 40s, there happened a change in his main interests: he turned to philosophy, specifically to the problem of the human soul. During that period, he published a few books on that subject, notably *Regnum*

Animale (*The Kingdom of the Anima* [soul]); volumes two and three were published about a decade later after he was shown a solution to the problem of defining the *soul*. We will say more about that later in this section.

His spiritual life started to unravel in the year 1743 when he was 55. It is not clear what exactly he experienced – perhaps it was a 'hallucination' – but whatever it was, it made a strong impression on him. So much so that his life changed profoundly, and he started to devote much more time to prayer and fasting. Here is a brief quote from his diary:[2] 'The whole day on April 9 [1744] I spent in prayer, in songs of praise, in reading God's Word [the Bible], and fasting.'

And then it happened: in April 1745 a man appeared to him in a vision while he was in an inn after dinner. Quoting from Swedenborg's letter to his friend Robsahm:[3]

> I went home, and during the night the same man revealed Himself to me again, but I was not frightened now. He then said that He was the Lord God, [], and that he had chosen me to explain to men the spiritual sense of the Scriptures.

Eventually, Swedenborg was shown much more than an interpretation of the scriptures. He was regularly taken to guided OB (out-of-body) tours in the realms of spirits, or in heaven; at times his OB travels lasted a few days of earthly time, non-stop; his guides were often perceived as angels. He was instructed to convey his experiences, and he dutifully wrote huge volumes of detailed spiritual travelogues of his 'thousands' of OB adventures.

In the scheme of the universe explained to him, the material world is surrounded by the world of spirits. Our existence in the former is in many ways inferior to the one in the latter. He writes:[4] 'The physical nature appended to us [] in the [] material world is not the real person but only the tool of our spirit.'

According to Swedenborg, death shifts people's primary consciousness from the material world into the timeless and spaceless world of spirits. Timelessness and spacelessness are some of the main features of the final mystic union, which we discussed earlier. At times Swedenborg is rather explicit about this; quoting from his book *Heaven and Hell*:[5] 'angels'

---

2 [Tro], p. 90.
3 Ibid, p. 95.
4 [Swe1], p. 334.
5 [Swe1], p. 221.

thoughts are not bounded [or] constrained by concepts drawn from space and time the way ours are'. One more quote on the same subject, this time from *Arcana Celestia (The Heavenly Arcana)*: 'Man should know that in their first cause and origin, the spaces and distances, and consequently the progression [time], which appear in the spiritual world, are changes of the state of the interiors [consciousness].'

Swedenborg tells us that there is an intermediate world between the material world and the world of spirits, where people go through various stages.

Communication in the spirit worlds of Swedenborg's travels was by telepathy; the word 'telepathy' was invented about a century later. He writes:[6] 'In heaven people actually speak directly from their thought, [...] a kind of audible thought.'

A noteworthy feature of Swedenborg's accounts in the present context is that some of the *people* he talked to were from other galaxies. Swedenborg does not identify which galaxies they were from; he just wrote that they were from 'earths in the starry heaven'. The only exception in his reports is a planetary solar system, the central sun of which is visible to us 'not far from the celestial equator'.

Not surprisingly, the notion of travel within the worlds of spirits is not ordinary (from *The Heavenly Arcana*):[7]

> To be conducted to the earths in the universe is not to be conducted and transported there as to the body, but as to the spirit; and the spirit is not conducted through spaces, but through variations of the state of the interior life, which appear to him like progressions through space.

In other words, space travel in the spirit world is equivalent to travel within one's consciousness.

The whole package of experiences delivered to Swedenborg seems to have been custom-suited to fit his social background. For example, whenever Swedenborg talked to the angels 'face to face', he went into their houses, which were just like the houses on Earth except more beautiful! Spousal relationships also transferred to heaven, and so did the traditional roles of husband and wife. Quoting from *Heaven and Hell*:[8] 'In heaven the

---

[6] Ibid, p. 90.
[7] [Swe2], p. 445.
[8] [Swe1], p. 294.

husband plays the role labeled intellect and the wife the role called volition
(the ability to make conscious choices and decisions).' Further:[9]

> The fact is that after death a man is none the less a man; and so
> fully is he a man that he does not know but that he is still living in
> the former world. He sees, hears and speaks; he walks, runs and
> sits; he lies down, sleeps and wakes, he eats and drinks; he enjoys
> the delights of married life [!], as he did in the former world; in a
> word, he is in all respects a man.

Somewhat unexpectedly, Swedenborg did not have an adverse attitude
toward sexual desires:[10] 'As for sexual love, it is the universal love, having
been put by creation in man's very soul, which is his whole essence.'

Angels, as it seems, went one step further:[11] '[Marriage in heaven] is
called "living together" [and] two [angelic] spouses in heaven are not
called two angels but one!'

He had a direct experience of what it meant to be 'angelic':[12]

> I have sometimes been let into the state in which angels are, and in
> that state I have spoken with them; and then I understood all; but
> when I was brought back to my former state, and thus into the natu-
> ral thought proper to man, and wished to recollect what I had heard,
> I could not; for there were thousands of things not adapted to the
> ideas of natural thought, thus not expressible at all by human words,
> but only by variegations [light-color changes] of heavenly light.

> (FROM *HEAVEN AND HELL*)

Here is a selection of Swedenborg's wisdom:

1. *The human soul [...] is immortal in all respects.* (*Heaven and Hell*)

2. *Thought is nothing but internal sight.* (*Divine Love and Wisdom*)

3. *Without love there is no life, and the life is of such a quality as is the
   love.* (*Divine Love and Wisdom*)

---

[9] [Dus], p. 72.
[10] [Dus], p. 110; the original source is Swedenborg's *Divine Love and Wisdom*.
[11] [Swel], p. 294.
[12] [Dus], p. 133; the original source is Swedenborg's *Heaven and Hell*.

Lest one dismisses out of hand Swedenborg's accounts of his spiritual travels as fantasies of a delusional individual, we note that there are some well-
authenticated examples of his clairvoyance. For example, on July 19, 1759,
while he was attending a social event in Gothenburg, on the west coast
of Sweden, Swedenborg suddenly became distressed: he *saw*, through a
kind of clairvoyant distant viewing, that there was a fire in Stockholm,
hundreds of miles away, and he provided a kind of live coverage of the
event. He was particularly distressed because his house was in the path
of the fire. His clairvoyant transmission of the event included details (for
example, that the fire was extinguished a few meters from his own house)
and was directly witnessed by 15 persons and indirectly by many more,
including the governor of Gothenburg, who called Swedenborg later that
day. It was days later that the conflagration was confirmed by couriers
dispatched to Stockholm. During the Great Stockholm Fire of 1759, some
300 houses were burnt down to ashes.

There were a few instances when, in order to gain access to ostensibly inaccessible information, Swedenborg simply asked his acquaintances
from the spirit world! One such authenticated example was the case of
Madame Herteville, the widow of the Dutch Ambassador in Stockholm,
who was asked by a goldsmith to pay for a service that her late husband had
purchased but had not paid for. The rest of the story comes from Emanuel
Kant's book *Dreams of a Spirit Seer*.[13]

> The widow was convinced that her late husband had been much
> too precise and orderly not to have paid this dent, yet she was
> unable to find this receipt. [So, she kindly asked Swedenborg to
> get in touch with her deceased husband…] Three days afterwards
> the said lady had company at her house for coffee, Swedenborg
> called and in his cool way informed her that he had conversed
> with her husband. The debt had been paid several months before
> his decease, and the receipt was in a bureau in the room upstairs.
> The lady replied that the bureau had been quite cleared out and
> that the receipt was not found among all the papers. Swedenborg
> said that her husband had described to him, how, after pulling
> out the lefthand drawer a board would appear, which required
> to be drawn out when a secret compartment would be disclosed,
> containing his private Dutch correspondence. [The whole party

---

[13] [Lod], p. 121.

then went upstairs and found the receipt in the bureau as told by Swedenborg. Nobody, of course, knew of the secret compartment.]

Swedenborg is described by his friends, and by those who viewed his experiences with deep skepticism – in other words, by everyone who got in contact with him – as being of virtually impeccable character:[14] 'no vehemence, anger, nor hatred; no sarcasm, contempt, nor fretfulness.' Count Höpken wrote in a letter in May 1773 that[15] 'The late Swedenborg certainly was a pattern of sincerity, virtue, and piety.'

The length of his spiritual travels increased with Swedenborg's age. Not long before his death he (quoting Robsahm),[16] 'lay some weeks in a trance, without any sustenance; and came to himself again.'

The Old Testament notwithstanding (Ecclesiastes 9:12), Swedenborg knew the date and time of his death: he told[17] John Wesley (the founder of Methodism) that he could not meet him at a certain time six months later, since he was to die 'on twenty-ninth next month'; which is what happened.

The following account of his death is by a minister named Ferelius, who administered the last rites:[18]

> I observed to him, that, as quite a number of people thought his sole purpose in promulgating his new theological system has been to make himself a name, or to acquire celebrity []. He thereupon half rose in his bed, and laying his sound hand [he was half-paralyzed by a stroke] upon his breast said with some manifestation of zeal: 'As true as you see me before your eyes, so true is everything that I have written, and I could have said more, had it been permitted.

There is a law based on the principle, *nemo moritimus preasumitur mentiri*, i.e., 'a dying person is not presumed to lie', that apparently may be in effect even today!

Swedenborg seemed pleased by his pending death,[19] 'as if he was going to a holiday'! He died on March 29, 1772.

---

[14] [Tro], p. 277.
[15] Ibid, p. 278.
[16] Ibid, p. 270.
[17] [Dus], p. ~155.
[18] [Dus], p. ~155.
[19] Ibid.

14.2 JOHANNES KEPLER: THE MYSTIC ASTRONOMER AND MATHEMATICIAN

Johannes Kepler (1571–1630) went down in history as a great astrono-mer, the originator of the (now-called) Kepler laws of planetary motion. However, he was primarily a mathematician: most of his life he was employed as such, mathematics played a crucial role in his research, and he had the mind of a mathematician. Moreover, and perhaps most impor-tantly, he considered himself a mathematician.[20]

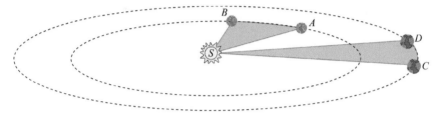

Kepler's first law on planetary motion: *The orbit of every planet is an ellipse, and the sun is in one of the foci.*

Kepler's second law on planetary motion: *If one planet moves from the position A to position B along its orbit, and if, during the same period of time, another planet moves from position C to position D, then the areas of the sectors SAB and SCD are equal.*

Johannes Kepler

---

[20] [Casp], p. 304.

Johannes Kepler was born in Weil (today Weil der Stadt), Germany on December 27, 1571. He was described as dark-eyed and black-haired, and he grew up to be 'small and gracefully built'. His father was a soldier, and his mother often followed her husband during his campaigns; Johannes was mostly abandoned to the care of his grandparents.

He was raised as a Catholic. This was an important aspect of his background, as during his times there were numerous conflicts and wars between adherents of various Christian denominations, mainly between Protestants and Catholics, in or around German-speaking lands.

In 1584, at the age of 13, Kepler entered the convent school, Adelberg. Four years later, in 1588, he enrolled in the University of Tübingen. His goal was to study theology, and the knowledge that he acquired, including mathematics and astronomy, was intended as preparation for such studies. However, as it happened, in the year 1594, when he finished his theological studies, he was offered the position of mathematics teacher in Graz. Kepler accepted and stayed there until 1600.

As a mathematics teacher, he was not very popular with the sons of nobles; he had very few students in the first year, and zero in the second. His patrons were rather tolerant and didn't blame him; they understood that mathematics 'was not everybody's meat'.

Kepler married in 1597. A son was born in 1598; he lived 60 days. His daughter, who came the following year, lived 35 days. He had four more children from his first marriage, two of whom survived to adulthood. These cold numbers hide tragedies. Especially heartbreaking for both parents was the death in 1611 of their six-year-old son Frederick. The mother couldn't recover and died soon after.

As a Catholic among Protestants, he was eventually forced to move out of Graz. In October 1600, he settled in Prague. Bohemia was also Protestant, but the emperor Rudolph was more tolerant and put art and science above religious narrow-mindedness. The next year Kepler was appointed as Imperial Mathematician in the court of the emperor. He stayed in Prague till 1612, the year Rudolph died.

From 1612 to 1626, he was the district mathematician in Linz. He married again in Linz; he chose his wife from no less than 11 candidates! The second wife, who was 24 when they married in 1613 (he was 42), eventually 'presented her husband' with six children, the first three of which died early. The last three of his children were born in 1621 (daughter), 1623 (son), and 1625 (son).

Kepler suffered from various ailments that affected his stomach and gallbladder. He was water-shy and rarely took baths and washed himself.

Between 1526 and 1527, he resided in Ulm, then moved to Regensburg. By the end of his career, he was the Court Astrologer to the Duke of Wallenstein. In the last few years of his life, he took numerous relatively short trips. He got sick during the last journey and died on November 15, 1530. His tombstone has the following distich that Kepler wrote for an epitaph:[21]

> I used to measure the heavens,
> now I shall measure the shadows of the earth.
> Although my soul was from heaven,
> the shadow of my body lies here.

Kepler's mysticism has its roots in his mother's belief in magical and mystical forces. Somewhat unwisely, she acted upon her beliefs – a dangerous pursuit at the times when witch-hunts were raging all over Germany. For example,[22]

> It was recalled that, many years before, [she] had asked the grave digger in her birthplace [] for the scull [of her dead father] in order to mount it in silver and have a drinking vessel made out of it for her son Johannes.

Her plans did not come to pass; the grave digger insisted on permission from the authorities, and the whole thing became too complicated to execute. Some years later, perhaps inevitably, she was accused of practicing witchcraft, arrested, and jailed. Mostly by an intervention of Kepler himself, she was spared torture and saved from being burnt at the stake; she was released after 14 months of imprisonment. She died the following year.

Kepler's mysticism manifested mainly in two areas of his pursuits: mathematics and astrology.

His mathematical mysticism was based on his strong belief that geometry played a crucial role in the divine creation of the universe. He wrote[23] in *Harmonices Mundi*, 1619: 'Geometry existed before the creation of things, as eternal as the spirit of God', and, further, 'Geometry is one and eternal, a reflection out of the mind of God'. So, it's not surprising that he turned to geometry when looking for an answer to the question that consumed him in 1595: 'Why are there six [!] planets?'[24] On July 19, he found

---

[21] [Casp], p. 370.

[22] Ibid, p. 251.

[23] [Roo], p. 623.

[24] [Casp], p. 66.

his eureka answer: because the solar system was created in a scheme based on the five Platonic solids (the Platonic solids are shown in Chapter 13)! Quoting from Casper, page 66: 'He was extremely excited. He believed that he had lifted the veil which hides the majesty of God and given a glimpse into his profound glory. The experience loosed a flood of tears'. In 1597, he published his discovery in his book *Mysterium Cosmographicum* (*The Mystery of the Universe*). The full title is quite a mouthful: *Prodromus Dissertationum Cosmographicum continents Mysterium Cosmographicum de admirabili Propertione Orbium Coelestium deque Causis Coelorum numeri, magnitudinis, motuumique periodicorum genuinis et propiis, demonstratum per quinque regularia corpora Geometrica.*

Kepler wanted to make a wooden model of his universe, but the craftsmen were not up to the task. So, we are left with his depiction of this beautiful and intriguing model of the solar system:

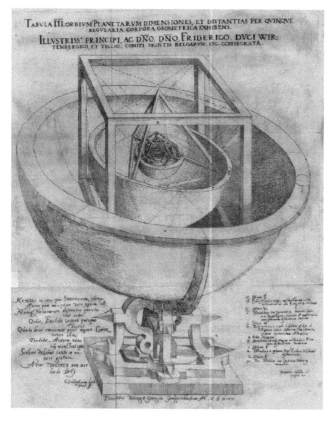

Kepler's model of the relative orbits of the first six planets of the solar system: the smallest sphere is inscribed in an octagon, which is in turn inscribed in a sphere, inscribed in an icosahedron, inscribed in a sphere, inscribed in a dodecahedron, inscribed in a sphere, inscribed in a tetrahedron, inscribed in a sphere, inscribed in a cube, inscribed in a sphere. The radii of the six spheres give approximations of the relative orbital radii of the first six planets, from Mercury to Saturn, the sixth and the last *known* planet in Kepler's time.

Kepler was neither the first nor the last to use Platonic solids in his models of the universe. In the scheme of the world imagined by Plato himself, the Earth, being the least mobile, was represented by a cube, and then, in the order given by their mobility, water was represented by an icosahedron, air by an octahedron, and fire, the most mobile, by a tetrahedron. According to Plato, God used the fifth regular polyhedron (the dodecahedron) 'for embroidering the constellations of the whole heaven'. Centuries after Kepler, Leadbeater, the clairvoyant theosophist mentioned in Chapter 10, *saw* that the elementary units of matter were distributed within the atoms at the vertices of the Platonic solids, and of some other related solids.[25]

Kepler had been an ardent astrologist since his student years; he was a popular calendar maker then, as well as during his Linz years. 'Calendar' meant something else during Kepler's times: 'The calendar men were expected to furnish information about the weather and the harvest prospects, about the war [], about religious and political events.'[26] All of this was based on astrological readings. Kepler composed six rather successful and sought-after calendars between 1617 and 1624 in Linz. Apparently, he scored at least some hits with his astrological guesses. However, Kepler was not an indiscriminate believer in astrological divinations. He wrote: 'It is also not credible that it can be seen from the horoscope how things will work out for anybody with certainty. In general, everybody is the master of his fate.'

Not surprisingly, geometry played a prominent role in Kepler's astrological readings too. According to him, 'the heavenly bodies travelling along the zodiac produced an excitement within our souls whenever they formed angles corresponding to those of the regular polygons.'[27]

Kepler's theory on the mode of interaction between planets and humans is pure mysticism: according to him, our planet was an ensouled, sentient entity! He wrote:[28] 'The natural philosophers may say what they want, there exists in the Earth also a soul. [] ['T']here is in the Earth [] an intelligent soul.' Since postulating the same for the other *heavenly bodies* is but a minor extension of this claim, all the major tenets of astrology become a kind of corollaries because interactions between planets and humans can then be reduced to contacts between intelligent beings!

At least one of Kepler's astrological claims has been checked statistically! Kepler wrote:[29] 'There is one perfectly clear argument beyond all

---

[25] Leadbeater: *Occult Chemistry*.
[26] [Casp], p. 61.
[27] [Sey], p. 70.
[28] [Casp], p. 99.
[29] [Sey]. p. 134.

exception in favour of the authenticity of astrology. This is the common horoscopic connection between parents and children.' The researcher Michel Gauquelin, mentioned in Chapter 13, decided to precisely investigate this claim. He found out that if both parents were born when Venus, or Mars, or the Moon, was just rising or near the highest point in the sky, then it was likely that the respective celestial body would be at the same relative position at the birth of their children.

Summarizing: Kepler's mathematical and astronomical research was, by and large, intertwined with the mysticism of divine geometry. The philosophy of Kepler's astrology rested on the mystic thesis that the Earth, and the planets in general, are sentient, conscious entities.

14.3 JOHN DEE: THE MYSTIC PARANORMAL RESEARCHER

To some extent, John Dee's life parallels Emanuel Swedenborg's: both were relatively minor mathematicians and major mystics. However, their mysticisms were very different: Swedenborg's was experiential (or true mysticism), and Dee's was mostly experimental and oriented in the direction of exploring paranormal phenomena.

According to John Dee's horoscope, he was born in London on July 13, 1527. In 1542, he entered St. John's College, Cambridge, and graduated in 1546.

John Dee

The next year he traveled to the Netherlands and stayed there for a year. The main purpose of the trip was 'to confer with learned men of the Dutch

Universities upon the science of mathematics, to which he had already begun to devote his serious attention.'[30]

Dee returned to Cambridge in the year 1548, then traveled to what is now Belgium, where he enrolled at the University of Louvain. He graduated in two years; it has been presumed that the title 'doctor', usually given to him, was earned there. Among the people that he befriended during these two years was the famous cartographer Gerardus Mercator.

Dee was only 23, already with the reputation of an eminent mathematician, when he went to Paris. There he gave a series of lectures in Euclidean geometry. He was so popular that the large audience could not be accommodated in the largest classrooms, and there were students who climbed outside the windows of the school classroom from where they could only see the lecturer.

At that time, Dee was offered to become the King's Reader in Mathematics at Paris University, but he declined it. After he returned to England in 1551, he was invited to become a lecturer in mathematical science at Oxford, but he declined that too. His main interests shifted in an entirely different direction. Nevertheless, he was still considered a prominent mathematician throughout his life, and in 1570 he was asked to write the preface to the first English translation of *Euclid's Elements* by Henry Billingsley. Not surprisingly, it can be discerned from this preface that Dee's view on mathematics had a distinctive mystical flavor. Here are a few tidbits:[31]

1. All thinges do appeare to be Formed by the reason of Numbers. For this was the principall example or patterne in the minde of the Creator.

2. Musike is a Mathematicall Science.

3. Astrologie, Is an Arte Mathematicall.

Dee was soon to be employed by Queen Elizabeth, mainly as her astrologer, but also as a prospective alchemist. She remained his patron until she died in 1603. Elizabeth was very fond of him, and she called him 'my philosopher'.

---

[30] Charlotte Fell Smith: *John Dee*, p. 8.
[31] [Dee]*.

Dee married three times; the first two wives died soon after he married them. In 1578, he married for the third and final time. The bride, Jane, was 22; he was 51. It was a relatively happy union, and in due course they had eight children, starting with his first son who was born July 13, 1579, on John Dee's 52nd birthday. Jane was a protective and devoted mother, and six of the children survived to adulthood.

We back up in our chronology to mention that in 1552 Dee met with Cardano when the latter was on his way back from Scotland. Besides their mysticism and interest in the 'occult', there was also an idiosyncratic event that connected them: they both witnessed an appearance in the sky of what is now commonly called a UFO. Cardano's account was given in Section 13.3; here is how Dee described it:[32]

> a strange Meteore in forme of a white clowde crossing galaxium, lay north and sowth over our zenith. This clowde was at length from the S. E. to the S. W., sharp at both ends, and in the West it was forked for a while. It was about sixty degrees high, it lasted an howr [!], all the skye clere abowt and fayr star-shyne.

Meteors in the sky are visible for at most a few minutes.

In 1564 Dee published the esoteric work *Monas Hieroglyphica*. The manner in which he produced this book shows unambiguously that he was already far along mystic roads. In fact, it was delivered to him in only 12 days by *spirits* that he encountered on his mystic travels. Dee wrote:[33] '[I am] the pen merely of [God] Whose Spirit, quickly writing these things through me, I wish and hope to be.'

In 1581, a man who introduced himself as Edward Talbot entered Dee's life. Dee came to like Edward, and he appreciated his talents. So much so that Edward soon became a member of Dee's household. As it turned out, their close association lasted some ten years. Edward was a *medium* whose presence there manifested a variety of paranormal phenomena. He was also somewhat of a swindler; his real name was Edward Kelley. We will describe and discuss their experiments further below.

In 1583, a certain Polish nobleman, Prince Laski, who was in England for a few months, attended several mediumistic séances with the Dee-Kelley duo. He was so much impressed by the manifesting phenomena

---

[32] [Smi], p. 65.
[33] [Pet], p. 6.

that he invited Dee and Kelley, together with their whole families, for a visit to Poland. The invitation was eventually accepted, and Dee was off for a long trip.

The journey was supposed to last two years; however, it took eight years before Dee and his family, enlarged by a couple of children, returned to England. Besides Poland, they had long sojourns in Bohemia, Germany, and the Netherlands.

It has been alleged that during this long trip, Dee was a spy for Queen Elizabeth, with whom he maintained indirect contact through letters. He sometimes signed these confidential letters to the queen with 007, and it has been postulated that he was the prototype for the fictional character James Bond.

Kelley never returned. He worked his way into the court of King Rudolph of Bohemia, promoting himself as an alchemist. However, his swindling schemes were soon uncovered, and he was imprisoned for two and a half years for deception. He was released for a year, then imprisoned again. His life ended soon thereafter. The story is that he broke both legs and sustained other injuries attempting to escape by a turret window. He may have been killed.

Dee stayed in Cambridge for a few years, then moved to Manchester in 1596 where he was installed as a warden in Christ's College. He returned to Cambridge in 1602.

Queen Elizabeth died in 1603 after 53 years of reigning, and Dee's financial situation significantly worsened. His wife Jane, 30 years his junior, died in 1605. The last few years of his life he lived close to his oldest son; he enjoyed being with his grandchildren. His last two astrological horoscopes were drawn for the first two grandchildren. Dee died in 1608.

Dee's passion in what would today be somewhat derogatorily called the occult, and what was then referred to simply as magic, started early. He was still a teenager of 17 or 18 when he was arrested because a certain George Ferrys alleged that 'one of his children had been struck blind and another killed by Dee's magic'. Dee was eventually cleared and liberated.

Dee was an astrologist virtually all his life. Astrology was a reputable scientific discipline during Dee's times, on par with astronomy and mathematics. Every self-respecting king had a personal astrologer. Alchemy had a similar status. Dee carried out many alchemical experiments with great passion and expectations. However, it seems that he never reached realms much beyond mechanical manipulations, or what a couple of centuries later Newton called vulgar chemistry.

Dee can be described as an early researcher of paranormal phenomena. Most of his experiments were associated with the medium Edward Kelley, in whose presence there manifested entities that Dee identified as angels. Kelley was primarily a scryer or a crystal gazer. We must immediately point out that, during the majority of sittings, he was the only one who could see or hear the angels; to Dee, they were 'invisible to the eye and ear'.

There was a total of 49 angels who manifested, all introducing themselves with names. Dee diligently recorded the proceedings of the séances in his *Actio Saulina*, starting with the first one that took place on December 21, 1581, when Kelley transmitted a message from the angel Annael. Dee's transcripts were published in five books after his death. A good modern reference is in Joseph H. Peterson's *John Dee's Five Books of Mystery*.[34]

The *angels* generally appeared on the surface of a 'shew stone,' a dark polished stone, but occasionally they stepped out into a beam of light and moved about the room. During such instances, Dee could also perceive them.

Sometimes during the séances, there happened phenomena that resulted in tangible, material objects: apports. *Apports* are defined as objects materialized from, so to speak, thin air, or transported through *solid* matter. For example, during one early evening séance, a little child appeared, standing in the sunset sunbeams holding something[35] 'most bright, most clere and glorius, of the bigness of an egg.' Then the angel Michael bade Dee to 'go forward, take it up, and let no mortal hand touch it but thine own'.

There were only three occasions when apports materialized. During another séance, angel Michael gave Dee a ring with a seal. We read from Dee's first book of mysteries:[36] 'Then he [angel Michael] toke [took] a ring out of the flame of his sword. [] After that, he threw the ring on the [] table.' Dee was overjoyed.

Given Kelley's shady character, and since the apparitions that manifested in his presence were almost always visible and audible only to him, the whole affair is the easy prey of skeptical criticism, the 'simplest' explanation being that Kelley lied and that he used sleight-of-hand tricks in the cases of apports. The point we are making here is that this is not the only simple explanation: under the very reasonable and virtually proven

---

[34] [Pet].
[35] [Smi], p. 86–87.
[36] [Pet], p. 78.

hypothesis that consciousness can exist incorporeally, the at-face-value interpretation of the Kelley affair becomes a virtual corollary, and so it is both simple and reasonable. Moreover, and equally importantly, there exists a library of hundreds of published experiments, including laboratory experiments under strict conditions, directly authenticating the phenomenon of apports.

In Section 10.6, we briefly described some of the 200 experimental séances performed by psychologist Kenneth James Batcheldor in his laboratory, under controlled conditions, and involving randomly chosen sitters; the purpose of the experiments was to deliberately induce manifestations of levitation. Besides levitation, and to the bewilderment of both the experimenters and the subjects/sitters, something else happened during some of the experiments: apports![37] A few times small objects like matches and pebbles mysteriously 'dropped' into the room from nowhere! Even large stones fell! One of them was given to a London museum for analysis. Subsequently, Batcheldor got a bewildering note from museum officials saying that the rock had disappeared!

Stone materializing from thin air in a closed room is a classical poltergeist effect, and it has been described a myriad of times throughout many centuries. Poltergeist stones are accidental phenomena, and as such they are not subjects of the investigation of experimental sciences.

For inducing apports deliberately, the Chick condition (from Section 10.6) applies: a *medium* is necessary. *Many* reputable scientists have investigated mediums in relation to apports, at times obtaining extravagantly positive results. Orthodox science, however, shut in the bunker made of materialistic axioms, either ignored these results altogether or faked embarrassment for the 'naïve and gullible' renegade scientists who dared take a step out of the bunker.

We finish this short foray into the phenomena of apports with the extraordinary Icelandic physical medium Indridi Indridason (1883–1912) Indridason was investigated by some prominent Danish and Icelandic scientists of the time. Here is what happened during an 'apport séance', winter 1906–1907:[38]

One evening that winter, after Indridason had fallen into a trance, the controls told the sitters that they could transport an object from any house

---

[37] [Brow], p. 66.
[38] [Gis], p. 86; reported by Nielsson, 1922.

in Reykjavik through the walls and the roof, and then bring it onto the table in the séance room. Nielsson (1922 [reference given]) reports:

> After the medium had gone into trance and was therefore unconscious, we first selected the house where the object was to be brought from, to exclude the explanation that he [] had brought the object with him. We allowed the controls to choose between the house of a well–known medical doctor and the bishop's residence. The controls chose the house of the doctor because the medium had often visited the home of the bishop. Immediately afterwards we heard a very peculiar knocking sound, the like of which I had never heard before, nor have I since then. There followed a short pause, during which the controls informed us that they had now got the object out through the roof of doctor's house. After a pause the knocking sound was heard for the second time, and shortly thereafter a large bottle containing a few bird specimens [preserved] in spirit landed down on our table.
>
> The doctor was contacted and eventually it was shown that a relative of the doctor kept the bottle in a chest, where it was missing when they inquired after the séance.

We end our narrative about Dee with the following beautiful prayer appearing in the preface of his book *Mysteriorum* (in Latin):

> Racte sapere, et intelligere doceto me, (O rerum omnium Creator,) Nam Sapientia Tua, totum est, quod vovo: Da verbum tuum in ore meo, (O rerum omnium Creator,) et sapientuiam tuam in corde meo fige.
>
> Teach me to perceive and understand properly (O Creator of all things), for your wisdom is all I desire. Fix your words in my ear (O Creator of all things), and fix your wisdom in my heart.

14.4 OMAR KHAYYĀM: POLYMATH AND SUFI MYSTIC

Abu'l Fath Omar ibn Ibrāhīm Khayyām, known as Omar Khayyām, was a Persian mathematician, philosopher, poet, astronomer, and Sufi mystic who lived from 1048 to 1131.

In or around 1070, he composed a treatise on algebra where he solved some geometrical problems and enumerated some types of cubic equations. By the way, algebra (from *al-jabr*, meaning 'completion') as an independent discipline had been introduced some 250 years earlier by Muhammad Al-Khwārizmi, another Persian sage. Al-Khwārizmi wrote a treatise where he showed how to solve quadratic equations.

Here is one of the problems that Khayyām discussed in his book (refer to the figure below):

Find the point G on a circle such that $\dfrac{OA}{OC} = \dfrac{OD}{DB}$, where O is the center of the circle and $OCGD$ is a rectangle.

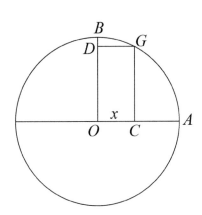

Omar Khayyām

Khayyām's solution involves circles, hyperbolas, and equations. It is rather long and complicated. However, we should keep in mind that he lived almost a millennium ago; even the (indispensable) equality sign = was introduced about half a millennium after Khayyām's times.

Here is a short solution to the problem using modern notation. We may assume that the radius of the circle is 1; denote $x = OC$.

Then $OA = 1$, $OD = \sqrt{1-x^2}$, $DB = 1 - \sqrt{1-x^2}$, so that $\dfrac{OA}{OC} = \dfrac{OD}{DB}$ turns into $\dfrac{1}{x} = \dfrac{\sqrt{1-x^2}}{1 - \sqrt{1-x^2}}$. Substituting $y = 1 - x^2$, the last equation reduces to the quartic equation $y^4 + 2y^2 - 4y + 1 = 0$. At this stage, we can use the formulas discovered by the Italian mathematicians of the 16th century, mentioned in the previous chapter. For the curious:

$$x = 1 - \frac{1}{9}\left(2 + \sqrt[3]{19 - 3\sqrt{33}} + \sqrt[3]{19 + 3\sqrt{33}} \right)^2 = 0.295598....$$

Omar Khayyām was born in 1048 in Nishapur, Persia, now Iran; Nishapur was then a cultural and trade center situated on the Silk Road.

Khayyām was tutored in mathematics from an early age by a disciple of the eminent philosopher, astronomer, mathematician, astrologist, alchemist, and writer Ibn Sina, known in the West as Avicenna, who he admired. In 1068, he moved to Samarkand, a scholarship center at the time. There he composed his treatise on algebra. In or around 1073, he led a group of scientists who were commissioned to revise the Persian calendar. He set up an observatory in Isfahan and eventually produced the *Jalali* calendar, the most precise calendar for many centuries (more accurate than the Gregorian Calendar of 1582, commonly used today).

Khayyām lived in the period of time when the Islamic Golden Age was coming to a close and when sultans were less and less tolerant of dissent and intellectual independence. Like many of the other polymaths that we have encountered in our narrative, he was careful to avoid antagonizing the religious orthodoxy. This is clear from his poetry:

The secrets which my book of love has bred,
Cannot be told for fear of loss of head;
Since none is fit to learn, or cares to know,
'Tis better all my thoughts remain unsaid.

The quote is from his large collection of quatrains (stanzas of four lines), called Rubaiyat in Persian. A selection of Khayyām's quatrains was translated[39] by Edward Fitzgerald in 1859 and is also called Rubaiyat. Most of the quoted poetry comes from Fitzgerald's selection.

One can also discern Khayyām's mysticism from his poetry; we see in the following quatrain the mystic doctrine that we have mentioned several times, according to which deep reality is hidden within oneself:

I sent my Soul through the Invisible,
Some letter of that After-life to spell:

---

[39] It is difficult to translate poetry: a literal translation will not convey its beauty, while the alternative often changes the intended meaning and inserts new metaphors. Take, for example, Khayyām's quatrain #69 at www.gutenberg.org/files/246/246-h/246-h.htm (Fifth Edition): Fitzgerald refers to chess in his translation. The French translation of the same quatrain utilizes the chess metaphor even more explicitly ([Kha]). I was about to add Khayyām to our curious thread of mathematicians who were also chess enthusiasts, but then, on second thought, I decided to check the original verse. I consulted two native Farsi speakers; both adamantly claimed that there is absolutely no reference to chess in the original Persian version.

And by and by my Soul return'd to me,
And answer'd 'I Myself am Heav'n and Hell.'

In the following quatrain, Khayyām seems to be alluding to the mystic union:

There was a Door to which I found no Key:
There was a Veil past which I could not see:

Some little Talk awhile of ME and THEE
There seemed—and then no more of THEE and ME.

The claim that Khayyām was a mystic is not based solely on his poetry. He was a Sufi, and Sufi's basic doctrine is fundamentally mystic. We will now say a few words about Sufism.

To some degree, Sufism is a mystic branch of Islam. According to current consensus, these two emerged concurrently. During its subsequent evolution, Sufism was influenced by Hinduism, Buddhism, and Zoroastrianism. The last one is a native Persian religious doctrine; Khayyām's father converted to Islam from Zoroastrianism.

Sufism is a practice rather than a theory, and its ultimate aim is entirely mystic: it is to achieve a 'union with God'. The initiation that sets a Sufi neophyte on that transcendent path may consist of long, rhythmic, synchronized chants performed in unison by groups of people.

Sufis call themselves 'Followers of the Real'. According to them, death is the ultimate means to achieve that reality and the expected mystic union. This credo can be recognized in the following verse by the great Sufi poet Jalāl ad-Dīn Rumi (1207–1273):

Death, that dread thing of which all mankind
stand in fear,
Is laughed and mocked at by saints when it
draws near.

Khayyām died in or around 1131.

CHAPTER 15

The Great Pythagoras

All the physical universe is not unique in nature and we must believe there are, in other regions of space, other words, other beings, other men.

LUCRETIUS (99–55 BC)

Pythagoras is probably most widely known for *Pythagorean Theorem*: $c^2 = a^2 + b^2$, with the notation established in the figure below. It is an elementary and fundamental theorem of cosmic universality. Regarding its universality, it is interesting to note that the illustrious Carl Friedrich Gauss tried to persuade the Russian sovereigns to mark out a huge replica of the figure we show below across the planes of Siberia, thus sending a signal to Mars, and possibly initiating interplanetary contact.

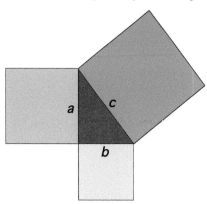

The Pythagorean Theorem: the sum of the areas of the squares with sides *a* and *b* is equal to the area of the square with side *c*

Pythagoras

DOI: 10.1201/9781003282198-15

It should then come as not surprising that what we now call the Pythagorean Theorem was known to the ancient civilizations of Egypt and Babylon thousands of years ago. In fact, Hecataeus of Aldera (around 360–290 BC) claimed that Pythagoras had learned this theorem from the Egyptians. In any case, Pythagorean greatness lies elsewhere. For example, according to Aristotle, Pythagoras believed in heliocentric theory[1]; he was some 2000 years ahead of his time.

15.1 PYTHAGORAS: A SHORT BIOGRAPHY

Pythagoras' lived in ancient times, about 2.5 millennia ago. There are no first-hand historical accounts of his life; the oldest sources are some 200 years after his death. So, some of the stories concerning his life are, perhaps rightfully, often classified as legends, which makes it difficult to extract the historic Pythagoras. The oldest preserved records regarding Pythagoras' life seem to be the treatises by Dicaearchus, 4th century BC, and the biography *Life of Pythagoras*, about 250 AD, written by Porphyry of Tyre,

Pythagoras was born around 570 BC on the island of Samos, off the coast of Asia Minor in the Aegean Sea. Samos was then an independent kingdom, and according to some sources, hosted a relatively advanced society. For example, during Pythagoras' times, a certain Eupalinos built a kilometer-long tunnel there by simultaneously drilling on both ends, meeting exactly at the middle.

When he reached adulthood, Pythagoras embarked on long travels that took him to Egypt and further to the east, possibly all the way to India. He stayed in Egypt for many years. He passed the tests for neophytes and was accepted as a disciple by the Egyptian priests and sages, who initiated him into ancient scientific and spiritual wisdom.

Upon returning to Samos, Pythagoras' reputation as an extraordinary teacher and a sage with miraculous mystical powers soon spread, and he attracted many followers. In due time he and his followers established new societies based on his teaching. These *Pythagorean Societies* soon started to flourish, and they spread all over the Hellenic world, from the west coast of Asia Minor, across the Islands of the Aegean Sea and mainland Greece, all the way to Southern Italy and Sicily.

Since the social conditions in his native Samos deteriorated, Pythagoras moved to the city of Crotona (or Croton) on the coast of Southern Italy, in the region populated by Greeks. At about this time the Pythagorean

---

[1] [Gui], p. 493–495.

schools and societies reached their peak, and many cities followed his social and spiritual doctrines. The domain of the Pythagorean world started steadily expanding northward, to the neighboring Italian tribes, including the Romans. According to a legend,[2] Numa Pompillus, the second king of Rome, was Pythagoras' disciple.

Another of his disciples, a young girl named Theano, became his wife. Pythagoras was sixty when he married. He fathered between three and seven children.

The following[3] description of Pythagoras is sourced in a description by Dicaearchus, 4th century BC:

> [Pythagoras] was very tall and of noble stature, and his voice, character, and every other aspect were marked by an exceptional degree of charm and embellishment. [] He dressed in a white robe, wore trousers (which was atypical for Greeks []) and crowned his head with a golden wreath []. [] Under his clothes too, there was something unique: the famous hard-to-interpret golden thighs with which Pythagoras was supposed to have been fitted.

The downward trend of the prosperity of the Pythagorean societies seems to have been triggered by the events related to a local aristocrat called Cylon. According to the story, supported by some ancient writings,[4] Cylon was of considerable influence, but of tyrannical temper. This Cylon wanted to become Pythagoras' student, but was refused by the great man, perhaps after some deliberation. The unfortunate outcome of this decision was Cylon's uprising that brought about the eventual demise of Pythagorean societies.

For all we know – and we don't know much – Pythagoras may have deferred his decision regarding Cylon to the caprice of Destiny by, say, flipping a coin from the highest cliff, thus creating a kind of quantum event. What would have happened if Pythagoras decided otherwise and accepted Cylon as his student? Was this the turning moment? Would a harmonious and universal society have branched from the alternative outcome? Remarkably, we have a tentative answer to that question! We will come back to this topic by the end of this chapter.

---

[2] [Rie], p. 12; there is some incongruence here: Numa is said to have lived a couple of centuries before Pythagoras.
[3] [Rie], p. 2.
[4] [Rie], p. 19.

It has been asserted that during the subsequent upheaval Pythagoras moved to Metapontum, to the north of Croton. Here is what Porphyry wrote[5] regarding Pythagoras' death:

> Others say that his companions, when fire was consuming the house in which they had just gathered, threw themselves into the fire and made a path for their teacher by using their own bodies to make a bridge over the fire. Pythagoras was despondent because he had lost his companions and took his own life.

This allegedly happened in 495 BC. After Pythagoras' death there were more rebellions against Pythagorean societies. This, together with the internal strife and antagonism that developed, eventually resulted in their complete demise, about 250 years after his death.

15.2 PYTHAGOREAN SOCIETIES, AND PYTHAGORAS' MYSTICISM

Even though Pythagoras coined the word *philosophy*, he was not as much of a philosopher as he was a great spiritual leader. He was a kind of holy man, the Western counterpart of Buddha. We know of the enormous impact of Buddha's teaching on the history of the Orient; Pythagoras' impact on the history of the Occident could have been proportional had the proverbial coin-tossup that Destiny is often fond of playing resulted with the other side up, and the Pythagorean societies had spread over all of Europe.

Coincidentally – or not – Pythagoras and Buddha were contemporaries: the former was only seven years older than the latter!

Here, in brief, is Pythagoras' program: The main goal of creating the Pythagorean societies was the attainment of '*harmonia*' (harmony), with emphasis on love/friendship and universal wisdom. A prerequisite to achieving this goal is an understanding of deeper reality. For this, one needs to understand oneself ('*the kingdom of heaven is within man*'), which in turn is achieved by applying practical mysticism. The ultimate goal is very mystic: purification in preparation for a return to the Supreme Mind.

Pythagoreans and their followers adhered to strict social and spiritual norms and procedures. They were strict vegetarians, with some further idiosyncrasies; for example, according to some sources, they abstained from consuming beans. Killing of nondestructive animals

---

[5] Ibid, p. 20.

and consumption of all animals was resolutely forbidden. According to the mathematician Eudoxus of Cnidos (~390–340 BC), Pythagoras even avoided contact with hunters.[6]

Becoming a true Pythagorean – as opposed to a Pythagorist, or a follower – was not an easy task: the applicant would first have to undergo three years of contempt, '*to test how [the applicant] was disposed to stability and true love of learning.*'[7] There followed five years of silence, with the aim of testing and developing self-control. After this the candidates became members of the inner circle, and they were allowed to receive Pythagoras' teaching directly from him. Pythagoras himself had a status between humans and God, which was reinforced by the miracles he performed.

The main Pythagorean spiritual doctrine was *Metempsychosis*, according to which the soul can exist apart from the body; we established earlier that there exists strong scientific support of the reasonability and rationality of this postulate. Within Metempsychosis, the main notion is *transmigration* of the soul: at death the soul continues existing incorporeally, and at the subsequent birth, it passes into another body. This body is not necessarily human; for example, it could be the body of an animal! On a related matter, according to the biographer Diogenes Laertius (about the 3rd century BC), Pythagoras retained '*memories of the events of ten and twenty generations*'[8] of his previous incarnations.

Pythagoras taught that the whole universe is animated by a great Soul. Like quite a few mathematicians that we encountered earlier, Pythagoras believed that the planets were soulful and intelligent. According to him, every material entity has body, soul, and spirit, and the last exists forever.

Pythagoras' mysticism extended to numbers: '*The first ten numbers are identified, on the basis of their structure and properties, with certain divinities and with ethical or physical concepts.*'[9] Unity, or the number 1, was called *monad*, the term he also used for the Supreme Mind or God. This was the original source of Leibniz's terminology.

15.3 PYTHAGORAS' MIRACLES

Now we take a look at Pythagoras' miracles.

There is the story about the Daunian bear, who (!) had the habit of killing livestock in the neighborhood. Pythagoras *talked* to the bear and

---

[6] [Rie], p. 37.
[7] Quote attributed to Iambilichus of Chalcis. See [Rie], p. 101.
[8] [Lon], p. 21.
[9] [Rie], p. 126.

convinced it not to harm any living creature. From that time the bear turned to vegetarianism and never attacked 'irrational animals' anymore. Pythagoras' power extended to birds too. According to Porphyry's biography *'he [Pythagoras] is said to have drawn an eagle which flew overhead, and after stroking it, he released it again.'*

There is a similar medieval account of a human–animal rapport. It comes from Adomnan's *Life of Columba*, a 7th century treatise on the life of St. Columba, and it is the first written description of the Loch Ness monster[10]:

[The monster] suddenly swam up to the surface, and with gapping mouth and with great roaring rushed towards the man swimming in the middle of the stream. [ ] The blessed man [Columba], who was watching, raised his holy hand and drew the saving sign of the cross in the empty air, and then, invoking the name of God, he commanded the savage beast, and said: 'You will go no further. Do not touch the man; turn backward speedily.' Then, hearing this command of the saint, the beast, as if pulled back with ropes, fled terrified in swift retreat; although it had before approached so close to [the man] as he swam that there was no more than the length of one short pole between man and beast.

Back to Pythagoras, the following curious story again comes from Porphyry's biography[11]:

Meeting some fishermen who were drawing their nets heavily laden with fishes from the deep, he [Pythagoras] predicted the exact number of fish they had caught. The fishermen said that if his estimate was accurate, they would do whatever he commanded. He then bade them to return the fish alive into the sea after having counted them accurately; and, what is more wonderful, not one of them died, although they had been out of the water for a considerable time.

Assuming that the last story is not apocryphal, a mystic might argue that Pythagoras received the correct number through an intervention by

---

[10] Quoted in [Nig].
[11] [Rie], p. 3.

(pure) *spirits*. In any case Pythagoras was very much interested in establishing contact with spirits. We are told[12] that Pythagoras conducted spiritistic séances with a table on wheels: the table freely moved toward signs and numbers, and Pythagoras interpreted the messages as revelations coming from an unseen world. The modern and more compact variant of Pythagoras' table on wheels is called a Ouija board.

We will pay more attention to the following outlandish episode, also coming from Porphyry's biography[13]:

> Almost unanimous is the report that on one and the same day he [Pythagoras] was present at both Metapontum in Italy and the Tauremenium in Sicily, in each place conversing with his friends, though the places are separated by many miles, both at sea and land, demanding a journey of great many days.

In modern terminology this is a case of *bilocation*. There are many recorded cases of such type, some happening spontaneously, some deliberately. The Catholic Church accepts some well-documented cases as true episodes of bilocation. For example, in September 1774, Alphonsus Liguori, a monk in Arezzo, fell into a five day long cataleptic trance. At the same time, and according to many witnesses, he was seen in Rome, a four day trip away, praying by dying Pope Clement's bedside. From the point of view of the Catholic Church this was a well-authenticated bilocation case.

We will now make an exception and recount an anecdote. In order to conform with our decision not to rely on anecdotes, we will designate it as being given 'for entertainment purposes only.' The following is a relatively recent amusing case involving the Italian mystic, Father Pio.[14]

> Once, just before going on the air, the announcer's [an Italian radio announcer] head ached violently that he was temporarily paralyzed. A few seconds later Padre Pio came into the studio, put his hand on the man's forehead, and the headache vanished. Astonished as the announcer was, he later decided that it must have been a hallucination. He went to see the priest to tell him what had happened, but before he could open his mouth, Pio put

[12] Melton, J. Gordon: *Encyclopedia of Occultism & Parapsychology*, 4th edition, 1996.
[13] [Rie], p. 4.
[14] [Gre], p. 75.

his hand on his visitor's forehead and said, smiling, 'Oh, oh, these hallucinations.'

From the point of view of materialistic science, cases of bilocation are not worthy attention, and the associated phenomena are dismissed out of hand. It is deemed unscientific to even consider the possibility of these events being anything else but pure fiction.

There seem to be no direct, scientific experiments with bilocation – which is hardly surprising, given the relative uncommonness of the phenomenon and of people who could deliberately induce it, and given the risks involved in merely considering the possibility of its genuineness, and then studying it (anything from scorn and ridicule to a ruined professional career). The closest we could get are the results of the laboratory-controlled experiments performed in the spring of 1973 by the psychologist Robert Morris with the psychic Keith (Stuart Blue) Harari. Harari managed to project a (holographic) image of himself into another room. These projections were observed by two beings: Harari's cat and a snake, kept in separate cages! The cat stopped meowing, and the snake bit and gnawed the cage at precisely the periods of time when Harari was attempting to project. This experiment may not be very convincing. We should note anyway that the visible spectrum of many animals is wider than that of humans'.

At the end of our narrative, we return to speculations: how would history have developed had Pythagoras preempted Cylon's rebellion by accepting Cylon's application to be his student? Amazingly we have a person who had experienced, in a way, this alternative history!

The following case comes from the files of the psychologist Kenneth Ring.[15] The consciousness of a 14-year-old girl split during the few minutes while she was drowning. The part of her consciousness that was out of her body was made aware of three lines of trajectories leaving from the past to three futures. We read from Ring's book: *Future A was a future that would have developed if certain events had not taken place around the time of Pythagoras three thousand years ago. It was a future of peace and harmony, marked by absence of religious wars.'* Futures B and C were destructive and disharmonious.

We are almost certainly experiencing a *wrong* future.

At the start of our journey, we mentioned that Grothendieck died in the year 2014. They say that Pythagoras was born in 570 BC.

---

[15] [Rin], p. 217.

Bibliography

[Acz]Amir D. Aczel: *Descartes Secret Notebook; A True Tale of Mathematics, Mysticism, and the Quest to Understand the Universe*, Broadway Books, New York, 2005.

[Aks]Alexander Aksakov: *Animism and Spiritism*, 1890 (in German), 1893 (in Russian, expanded and corrected), 2001 edition in Russian.

[Bab]Edwin Dwight Babbitt: *The Principles of Light and Color; Including Among Other Things the Harmonic Laws of the Universe, the Etherio-Atomic Philosophy of Force, Chromo Chemistry, Chromo Therapeutics, and the General Philosophy of the Fine Forces, Together with Numerous Discoveries and Practical Applications*, Babbitt & Co,, New York, 1878.

[Ball]W. W. Rouse Ball: *A Short Account of the History of Mathematics*, Dover, New York, 1960 (originally published 1908).

[Bat]*Kenneth James Batcheldor: Report on a Case of Table Levitation and Associated Phenomena, *Journal of the Society of Psychic Research*, **43**, 339–356, 1966.

[Bes]Annie Besant and Charles W. Leadbeater: *Occult Chemistry; Clairvoyant Observations on the Chemical Elements*, third edition, Theosophical Publishing House, London, 1919.

[Bis]Morris Bishop: *Blaise Pascal (A biography)*, Dell Publishing Co., New York, 1966.

[Bla]David Black: *Ekstasy; Out-of-the-Body Experiences*, The Bobbs-Merrill Company, Inc. Indianapolis and New York, 1975.

[Bon]Frederic Bligh Bond: *The Gate or Remembrance*, Blackwell, Oxford, 1918.

[Bos]D. N. Bose and Hiralal Haldar: *Tantras: Their Philosophy and Occult Secrets*, third edition, Firma KLM, Calcuta, 1981.

[Bro]* C. Brooks-Smith and D. W. Hunt: Some Experiments in Psychokinesis, *Journal of the Society of Psychical Research*, June, 40 (764), 1970.

[Brou]Luitzen Egbertus Jan Brouwer: Life, Art and Mysticism, *Notre Dame Journal of Formal Logic*, Vol. 37, No. 3, Summer 1996 (Walter P. van Stigt's translation).

[Bro1]Courtney Brown: *Cosmic Voyage; Astonishing New Evidence of Extraterrestrials Visiting Earth*, Onyx Book, New York, 1997.

[Bro2]Courtney Brown: *Cosmic Explorers; Scientific Remote Viewing, Extraterrestrials, and a Message for Mankind*, A Dutton Book, New York, 1999.

[Brow]Michael H. Brown: *A Report on the Powers of Psychokinesis, Mental Energy that Moves Matter*, Steinerbooks, Blauvelt, N.Y., 1976.

[BroR]Rosemary Brown: *Immortals by My Side*, Henry Regnery Company, Chicago, 1974.

[But]Johannes Von Buttlar: *Journey to Infinity; Travels in Time*, Collins, Fontana, 1975 translation of the 1973 German edition.

[Car]Jerome Cardan (Girolamo Cardano; Hieronimus Cardanus): *The Book of My Life (De Vita Propria Liber)*, 1575 (1931 English edition), J. M. Dent and Sons, London.

[Car2]Gerolamo Cardano: *Opera Omnia* (Introduction by August Buck), 1662 (1967 edition, Volume 1).

[Casp]Max Caspar: *Kepler* (translated from German by C. Doris Hellman), Collier Books, New York, 1962.

[Cas]Carlos Castaneda: *Journey to Ixtlan: The Lessons of Don Juan*, The Bronx, New York 1982.

[Cas2]Carlos Castaneda: *A Separate Reality: Further Conversations with Don Juan*, Simon & Schuster, 1971.

[Car]Hereward Carrington: *Psychic Oddities; Fantastic and Bizarre Events in the Life of a Psychical Researcher*, Rider & Company, 1952.

[Cha]Jean Jacques Charbonier: *7 Reasons to Believe in the Afterlife; A Doctor Reviews the Case for Consciousness After Death*, Inner Traditions, 2015 (French original published in 2012).

[Cou]Alyson Coudert, Richard Popkin and Gordon Weiner (editors): *Leibniz, Mysticism and Religion*, Kluwer Academic Publishing, 1998.

[Cum]Geraldine Cummins and E. B. Gibbes: *The Road to Immortality; Being a description of the After-life purporting to be communicated by the late F. W. H. Myers through Geraldine Cummings (With evidence of the survival of human personality by E. B. Gibbes)*, Ivor Nicholson & Watson, Ltd., London, 1932.

[Cur]Ian Currie: *You Cannot Die; The Incredible Findings of a Century of Research on Death*, Somerville House, Toronto, 1998.

[Dal]Dirk van Dalen: *Mystic, Geometer, and Intuitionist; The Life of L. E. J. Brouwer, Volume 1*, Clarendon Press, Oxford, 1999.

[Dal2]Dirk van Dalen: *Mystic, Geometer, and Intuitionist; The Life of L. E. J. Brouwer, Volume 2*, Clarendon Press, Oxford, 1999.

[Dan]Tobias Dantzig: *Henri Poincaré*, Greenwood Press, New York, 1954 (1968 reprint).

[Dau]Joseph Warren Dauben: *Georg Cantor*, Princeton University Press, Princeton, 1979.

[Daw]*John W. Dawson, Jr., The Published Work of Kurt Gödel: an Annotated Bibliography, *Notre Dame Journal of Formal Logic*, Vol. 24(2), 255–284, April 1983.

[Dee]*John Dee: Mathematical Preface to Euclid; can be found at http://www.mirrorservice.org/sites/ftp.ibiblio.org/pub/docs/books/gutenberg/2/2/0/6/22062/22062-h/22062-h.htm.

[DeM]C.D. Sophia Elizabeth de Morgan: *From Matter to Spirit: Ten Years Experience in Spirit Manifestations*, Cambridge University Press, 1863.

[DeM2]Sophia Elizabeth de Morgan: *Memoir of Augustus de Morgan*, Cambridge University Press, 1882.

[Dob]Betty Jo Teeter Dobbs: *The Janus Faces of Genius; the Role of Alchemy in Newton's Thought*, Cambridge University Press, 1991.

[Dos]Larry Dossey: *Healing Words; the Power of Prayer and the Practice of Medicine*, Harper, San Francisco, 1993.

[Dup]Yvonne Duplessis: *Paranormal Perception of Color*, Parapsychology Foundation, Inc., New York, 1975, Translated from French by Paul von Yoal.

[Dus]Wilson van Dusen: *The Presence of Other Worlds; The Psychological/ Spiritual Findings of Emanuel Swedenborg.* Harper and Row, New York, 1974.

[Fra]Marie-Louise von Franz: *On Dreams and Death, A Jungian Interpretation*, Shambhala, Boston, 1987.

[Gar]Martin Gardner: *Science; Good, Bad and Bogus*, Prometheus Books, Buffalo, New York, 1981.

[Gis]Loftur R. Gissurarson and Erlandur Haraldson: *The Icelandic Physical Medium Indridi Indridason*, in the *Proceedings of the Society for Psychical Research*, Volume 57, Part 214, January 1989.

[Göd]Kurt Gödel: *Collected Works: Volume IV: Selected Correspondence*, A-G, Oxford, 2006.

[Gre]Herbert B. Greenhouse: *The Astral Journey*, Avon Books, New York, 1974.

[Gro]Stanislav Grof: *The Cosmic Game: Explorations of the Frontiers of Human Consciousness*, State University of New York Press, New York, 1998.

[Gro2]Stanislav Grof, *When the Impossible Happens, Adventures in Non-ordinary Realities*, Sounds True, Louisville, CO, 2006.

[Gros]Michael Grosso: *The Man Who Could Fly; St. Joseph of Copertino and the Mystery of Levitation*, Rowman and Littlefield, Lanham, MD, 2016.

[Grot]Alexander Grothendieck: *Récoltes at Semailles, Part 1, The life of a mathematician; Reflections and Bearing Witness* (Translated by Roy Lisker, 2002 and on; web page).

[Gur]George Ivanovich Gurdjieff: *Meetings with Remarkable Men*, Picador/Pan Macmillan, London, 1978.

[Gui]Rosemary Elllen Guiley: *Encyclopedia of Mystical and Paranormal Experience*, Castle Books, Edison, NJ, 1991 .

[Hall]Manly P. Hall: *The Secret teaching of All Ages*, Philosophical Research Society, New York, 2003 (First published 1928).

[Hap]F. C. Happold: *Mysticism: A Study and an Anthology*, Penguin Books Ltd., London, 1972 edition (first published in 1963).

[Hil]Peter Hilton: *Algebra and Logic*, Springer Mathematical Notes 450; ed. J. Crossley, 1975.

[Hod]Andrew Hodges: *Alan Turing: The Enigma*, Princeton University Press, Princeton, 1983.

[Iind]Shotaro Iida: *Reason and Emptiness: A Study in Logic and Mysticism*, Hokuseido Press, Tokyo, 1980.

[Joh]George Lindsay Johnson: *Does Man Survive? The Great Problem of the Life Hereafter and the Evidence for its Solution*, Harper and Brothers, New York and London, 1936.

[Kan]Robert Kanigel: *The Man Who Knew Infinity; A Life of the Genius Ramanujan*, Charles Scribner's Sons, New York, 1991.

[Kee]S. V. Keeling: *Descartes*, Oxford University Press, Oxford, 1968.

[Ker]C. Louis Kervran: *Biological Transmutations*, Happiness Press, 1966 (1972 English translation).

[Kha]Omar Khayyam: *Rubaiyat*, translated in Farsi, English, German, French and Arabic, Arena of Persian Art & Thought, 2012.

[Kli]Jon Klimo, *Channeling, Investigations on Receiving Information from Paranormal Sources*, J. P. Tarcher, Los Angeles, 1987.

[Las]Ervin Laszlo (with Antony Peake): *The Immortal Mind; Sceince and the Continuity of Consciousness beyond the Brain*, Inner Traditions, Rochester, VT, 2014.

[Lei]Gottfried Wilhelm Leibniz: *Philosophical Papers and Letters*, D Reidel Publishing company, Holland, 1969 (First edition 1956).

[Lei2]Baron Gottfried Wilhelm von Leibniz: *Basic Writings; Discourse on Metaphysics, Correspondence with Arnauld*, Monadology, The Open Court Publishing Company, La Salle, 1968.

[Lis]*Roy Lisker: *Visiting Alexandre Grothendieck, June 1988*, http://matematicas .unex.es/~navarro/res/liskerquest.pdf .

[Lod]Sir Oliver Lodge: *The Survival of Man; A Study in Unrecognized Human Faculty*, George H. Doran, New York, 1909, 1920 edition.

[Lod2]Sir Oliver Lodge: *Raymond, or Life and Death, 1916*, ninth edition, Metthuen & Co., London, 1918.

[Lod3]Sir Oliver Lodge: *Why I Believe in Personal Immortality*, Doubleday, Doran & Company, Inc., Garden City and New York, 192969. [LodP]Paul Lodge (editor): *Leibniz and His Correspondents*, Cambridge University Press, Cambridge, 2004.

[Lon]Herbert Strainge Long: *A Study of the Doctrine of Metempsychosis in Greece; From Pythagoras to Plato*, Princeton University Press, Princeton, 1949.

[Lor]David Lorimer: *Survival; Body Mind and Death in the Light of Psychic Experience*, Routledge & Kegan Paul, London, 1984.

[Los]Mary Losure: *Isaac the Alchemist, Secrets of Isaac Newton, Revealed*, Candlewick Press, Somerville, Massachusetts, 2017.

[Low]Victor Lowe: *Alfred North Whitehead: The Man and His Work, Volume 1*, John Hopkins University Press, Baltimore, 1990.

[Low2]Victor Lowe: *Alfred North Whitehead: The Man and His Work, Volume 2*, John Hopkins University Press, Baltimore, 1990.

[Mac]Andrew MacKenzie: *Hauntings and Apparitions*, Heinemann, London, 1982.

[May]Elizabeth Lloyd Mayer: *Extraordinary Knowing: Science, Skepticism, and The Inexplicable Powers of the Human Mind*, Bantam, 2007.

[McC]Simon McCarthy-Jones: *Hearing Voices; The History and Meaning of Auditory Verbal Hallucinations*, Cambridge University Press, Cambridge, 2012.

[More]Henry More: *The Immortality of the Soul*, Springer, Dordrecht, Netherlands, 1642 (1987 edition Edited by A. Jacob).

[Mor]Melvin Morse and Paul Perry: *Closer to the Light: Learning from the Near-Death Experiences of Children*, Villard Books, New York, 1990.

[Mor]* M. Morse, P. Castillo, D. Venecia, et al.: Childhood Near-Death Experiences. *ADJC* 140 (1986), 1110–1113.

[Mos]Thelma Moss: *The Probability of the Impossible; Scientific Discoveries and Explorations in the Psychic World*, Routledge & Kegan Paul, London, 1976.

[Nig]Joseph Nigg: *The Book of Fabulous Beasts; A Treasury of Writings from Ancient Times to the Present*, Oxford University Press, Oxford, 1999.

[New]* Romaine Newbold: *Subconscious Reasoning*, The Proceedings of the Society for Psychical Research, vol. XII, pages 11–20, 1897.

[Ore]Oystein Ore: *Cardano, the Gambling Scholar*, Dover Edition, New York, 1965 (originally published in 1953).

[Osi]Karlis Osis: *Deathbed Observations by Physicians and Nurses*, Parapsychology Foundation, Inc., New York, 1961.

[Ost]Sheila Ostrander and Lynn Schroeder: *Psychic Discoveries Behind the Iron Curtain*, Prentice Hall, New Jersey, 1971 fifth printing (first edition 1970).

[Pas]Blaise Pascal: *Pensées [Thoughts]*, 1660 (translated by W. F. Trotter).

[Pet]Joseph H. Peterson, (editor): *John Dee's Five Books of Mystery; Original Sourcebook of Enochian Magic*, Weiser Books, Boston, 2003.

[Pla]Guy Lyon Playfair: *Twin Telepathy*, Vega Books, London, 2002.

[Plo]Plotinus: *Works, Volumes 1-7, (About Year 303, 30 Years After His Death)*, 1966 edition, With An English Translation, Notes and Synopses by A. H. Armstrong, Harward University Press, London.

[Poi]Henri Poincare (Stephen Jay Gould, editor): *The Value of Science: Essential Writings of Henri Poincare*, Modern Library, New York, (Oct. 2 2001).

[Rad]Dean Radin: *Supernormal; Science, Yoga, and the Evidence for Extraordinary Psychic Abilities*, Deepak Chopra Books, New York, 2013.

[Rad2]Dean Radin: *Entangled Minds: Extrasensory Experiences in a Quantum Reality*, Simon & Schuster, New York, 2006.

[Rao]Ramakrishna Rao (editor): *Basic Research in Parapsychology*, McFarland & Co., Jefferson, NC, 2001.

[Rhi]Louisa Rhine, *ESP in Life and Lab: Tracing Hidden Channels*, MacMillan, New York, 1967.

[Rhi2]Louisa E. Rhine: *Mind Over Matter; psychokinesis: The Astonishing Story of the Scientific Experiments that Demonstrate the Power of the Will Over Matter.* Collier McMillan Publishers, London, 1970.

[Rhi3]Louisa Rhine, *The Invisible Picture: A Study of Psychic Experiences*, McFarland, Jefferson, NC, 1981.

[Ric]Bob Rickard and John Michell: *The Rough Guide to Unexplained Phenomena*, RoughGuides, London and New York, 2007.

[Rie]Christoph Riedweg: *Pythagoras*, Cornell University Press, Ithaca/London, 2002, 2005 translation from German to English.

[Rin]Kenneth Ring: *Heading Toward Omega: In Search of Meaning of the Near-Death Experience*, W. Morrow, New York, 1985.

[Rob1]Jane Roberts (with Robert Butts): *The Unknown Reality, Volume 2*, Prentice Hall, Englewood Cliffs, N.J., 1977.

[Rob2]Jane Roberts (with Robert Butts): *Seth Speaks (The Eternal Validity of Soul)*, Amber-Allen Publ., New World Library, Novata, California, 1994 (first published 1971).

[Roo]Alexander Roob: *Alchemy and Mysticism*, Taschen, Koln, 2001.

[Ryz]*Milan Ryzl and J. G. Pratt: A Further Confirmation of a Stabilized ESP Performance in a Selected Subject, *The Journal of Parapsychology*, Vol. 27, June 1963, pages 73–83.

[Sab]Michael B. Sabom: *Recollection of Death: A Medical Investigation*, Harper & Row, New York, 1982.

[Sch]Winfried Scharlau: *Who Is Alexander Grothendieck? A Biography, Part 3: Spirituality*. Translated from German by Melissa Schneps, Norderstedt.

[Sch]* Winfried Scharlau: *Who is Alexander Grothendieck*, Notices of the American Mathematical Society, Vol. 63(3), 242–267, September 2008.

[Schw1]Gary E. Schwartz (with William L. Simon): *The Afterlife Experiment*, Pocket Books, New York, 2002.

[Schw]Gary E. Schwartz: *The Sacred Promise; How Science is Discovering Spirit's Collaboration with Us in Our Daily Lives*, Atria Books, New York, 2011.

[Sey]Percy Seymour: *Astrology; The Evidence of Science*, Arcana, London, 1990 revision of 1988 publication.

[Smi]Charlotte Fell Smith: *John Dee (1527–1608)*, Constable and Company LTD, London, 1909.

[Smy]Frank Smyth and Roy Stemman: *Mysteries of Afterlife*, Ferguson Publishing Company, London, 1991.

[Soa]S. G. Soal and H. T. Bowden: *The Mind Readers; An Important Recent development in Telepathy*, Doubleday & Company, New York, 1960.

[Spo]John Spong (editor): *Proceedings of the Symposium on Consciousness and Survival*, Institute of Noetic Sciences, Sausalito, CA, 1987.

[Stan]David Standish: *Hollow Earth*, Da Capo Press, Cambridge, MA, 2006.

[Star]Jonathan Star: *The Inner Treasure; An Introduction to World's Sacred and Mystical Writings*, Tarcher/Putnam Books, New York, 1999.

[Str]Whitley Strieber and Jeffrey J. Kripal: *The Supernatural; A New Vision of the Unexplained*, Jeremy Tarcher / Penguin, New York, 2016.

[Stre]B. H. Streeter and A. J. Appasamy: *The Sadhu; A Study in Mysticism and Practical Religion*, MacMillan and Co., London, 1927.

[Sum]Montague Summers: *The Physical Phenomena of Mysticism*, Barnes & Nobles Inc., New York, 1950.

[Sus]Henry Suso: *Little Book of Eternal Wisdom and Little Book of Truth*, About 1330, Harper, New York, 1953.

[Swe1]Emanuel Swedenborg, *Heaven and Hell* (1758), Massachusetts New Church Union, Boston, 1889.

[Swe2]Emanuel Swedenborg: *Arcana Coelestia (the Heavenly Arcana)*, Volume XI, 9112–9973, J. Hodson, London, 1789–1816.

[Tag]Lynne McTaggart: *The Intention Experiment; Using Your Thoughts to Change Your Life and the World*, Atria Books, New York, 2007.

[Targ]Russell Targ and Keith Harary: *The Mind Race: Understanding and Using Psychic Abilities*, Random House, New York, 1984.

[Targ2]Russell Targ: *Limitless Mind; A Guide to Remote Viewing and Transformation of Consciousness*, New World Library, Novato, Callifornia, 2004.

[Tart]* Charles Tart: Psychophysiological Study of Out-of-the-Body Experiences, in a Selected Subject, Journal of the American Society for Psychical Research, 62, 3–27, 1968.

[Tart2]Charles Tart (editor): *Altered States of Consciousness*, John Wiley & Sons, New York, 1969 (1990 updated edition).

[Thi]Rüdiger Thiele: Georg Cantor, in *Mathematics and the Divine: A Historical Study*, editors T. Koetsier and L. Bergmans, Elsevier, Amsterdam, 2005.

[Tie]Richard Tieszen: *After Gödel; Platonism and Rationalism in Mathematics and Logic*, Oxford University Press, Oxford, 2011.

[Til]William A. Tiller, Walter E. Dibble and Michael J. Kohane: *Conscious Act of Creation; An Emergence of a New Physics*, Pavior Publishing, Walnut Creek, California, 2001.

[Tro]George Trobridge: *Swedenborg, Life and Teaching*, Swedenborg Foundation, New York, 1962, Fifth reprint of the fourth edition, 1935.

[Tur]*A. M. Turing: Computing Machinery and Intelligence, *Mind*, 59, 433–460 (1950).

[Ulm]Montague Ullman and Stanley Krippner (with Alan Vaughan): *Dream Telepathy: Experiments in Nocturnal ESP*, Macmillan Publishing Company, New York, 1973.

[Vas]L. L. Vasiliev: *Experiments in Distant Influence; Discoveries by Russia's Foremost Parapsychologist*, Wildwood House Ltd., London, 1976 edition of 1963 transltion of 1962 original.

[Wan]Hao Wang: *Reflections on Kurt Gödel*, A Bradford Book, London, 1988.

[Wan2]Hao Wang: *A Logical Journey; From Gödel to Philosophy*, The MIT Press, Cambridge and London, 1996.

[White]John White (editor): *Psychic Exploration; A Challenge for Science*, Capricorn books, New York, 1976.

[Whi]Alfred North Whitehead: *Process and Reality; An Essay in Cosmology*, Free Press, New York, 1929.

[Whi2]Alfred North Whitehead: *Modes of Thought*, Macmillan, New York, 1938.

[Whm]J. H. M. Whiteman: *The Mystical Life*, Faber and Faber, London, 1961.

[Wie]Philip P Wiener (editor): *Leibniz; Selections*, Charles Scribner's Sons, New York, 1951.

[Wil]Colin Wilson: *Afterlife; An Investigation of the Evidence of Life after Death*, Grafton Books, London, 1985.

[Woo]Barry A. Woodbridge: *Alfred North Whitehead: A Primary-Secondary Bibliography*, Bowling Green State University, Bowling Green, Ohio, 1977.

Index

Note: parenthetical M stands for mathematician. In general, in the parenthetical remarks we list some of the interests of the respective persons.

Rosicrucian Brotherhood 155–157
Rudolph, King of Bohemia 182, 189
Rumi (Sufi poet) 7, 27, 195
Russell, Bertrand (M) 12, 64–65, **67–70**, 78
   HBR (hypothetical Bernard Russell)
   68–69
Ryzl, Milan (psi-researcher) 57–58, 60

S

Sabom, Michael B. (NDE-researcher) 71
Saint-Germain, Count (alchemist,
    never-ager) 124
Samadhi 88
Schlitz, Marilyn (psi-researcher) 172
Schmeidler, Gertrude (psi-researcher) 171
Schmidt, Helmut (psi-researcher) 125
Schumann, Robert (composer) 101
Schwartz, Gary E. (psi-researcher, writer)
    101–102, 111
Schwartz, Laurent (M) 22
Set of all sets 78–79
Seymour, Percy (writer) 171
Shiva (Hindu god) 124
Sidgwick, Henry (psi-research) 63
Sidgwick, Eleanor-Nora (M) **62–64**
Singh, Sadhu Sundar (mystic) 145–146
Slade, Henry (medium) 131
Soal, Samuel (M) **60–62**
Sophia, Duchess 120, 154
Steiger, Brad (writer) 131
Štepanek, Pavel (clairvoyant) 57–58, 60
Strauss, Ernst Gabor (M) 35
String theory 125
Subatomic particles of matter 127
Sufism 195
Suso, Henry (mystic) 25
Swann, Ingo (psychic) 71, 79, 125
Swedenborg, Emanuel (M, mystic, OBEr)
    173–180

T

Tanaroff, Alexander-Sasha 20–22
Tantras 124
Targ, Russell (psi-research) 125
Tart, Charles (psi-research) 70, 79
Tartaglia (M) 160

Tartini, Giuseppe (composer) 101
Telepathy 9, 44, 52–54, 61, 109, 112
Theresa, of Avila (mystic) 130
Tiller, William (psi-researcher) 125
Tieszen, Richard (biographer) 37–38
Topology 82–83
Tsiolkovsky, Konstantin 53
Transfinite numbers 76
Transmigration 200
Transmutation 123, 125–126
Turing, Alan (M) **44–55**, 96
   chess 49
   decision problem (solution) 45
Turing machines 45–47
Tyrrell, George N.M. (psi-research) 60
Tzu, Lao (philosopher) 104

U

UFO (unidentified flying objects) 168, 188

V

Van Dalen, Dirk (biographer) 88, 90, 100
Van Eeden, Frederic (mystic) 83
Van Stigt Walter P. (translator) 88
Vassy, Zoltan 170
Vedas 30
Vision 2
Voltaire (philosopher) 124

W

Wallace, Alfred Russel (writer) 131
Wang, Hao (biographer) 33, 36–39,
    41–42
Weil, André (M) 22
Wesley, John (the founder of
    Methodism) 180
Whitehead, Alfred North (M,
    philosopher) **64–67**
Whiteman, Joseph (M, OBEr) **72–73**
Wolf, Fred Alan (physicist) 26

Z

Zener cards 54
Zölner, Johann (psi-researcher) 131